Lecture Notes in
Operations Research and
Mathematical Systems

Economics, Computer Science, Information and Control

Edited by M. Beckmann, Providence and H. P. Künzi, Zürich

36

M. J. P. Magill

Indiana University, Department of Economics,
Bloomington, Indiana

On a General Economic Theory of Motion

Springer-Verlag
Berlin · Heidelberg · New York 1970

Title No. 3785

"Those who have meditated on the beauty and utility in theoretical mechanics, of the general method of Lagrange—who have felt the power and dignity of that central dynamical theorem which he deduced, in the Mécanique Analytique. . . must feel that mathematical optics can only <u>then</u> attain a coordinate rank with mathematical mechanics. . . in beauty, power and harmony, when it shall possess an appropriate method, and become the unfolding of a central idea."

W. R. HAMILTON 1833

CONTENTS

* * * * *

Chapter I

On the Origin of the Ideas

1. The purpose of the present inquiry is to develop a General Economic Theory of Motion, inspired on the one hand by the economic theory of motion developed by RAMSEY, FISHER, VON NEUMANN and SAMUELSON, and on the other hand by the beautiful and powerful methods of analytical mechanics introduced by EULER, LAGRANGE and HAMILTON. One part of the theory has its origin in the well-developed science of mathematical physics; the other part has its origin in the emerging science of mathematical economics. Consider first the origin of the ideas in mathematical physics.

2. Physics was the first of the Sciences to develop a general theory of motion. Indeed physics had its origin in the study of motion. As in the development of every science, observation preceeded analysis: the motion of material bodies was observed for more than two thousand years before GALILEO (1638, 1) revealed the true principles that govern their motion. The laws of motion (of material bodies) were first formally stated by NEWTON, in his wonderful "Philosophiae Naturalis Principia Mathematica" (1687, 2) which transformed the theory of motion into a new science.

In the century that followed EULER and LAGRANGE (1744, 3; 1788, 4) transformed the theory of motion into an even more powerful and elegant form. They showed that all problems of motion in mechanics could be unified by one <u>Central Principle</u>, the <u>Principle</u>

of Least Action, and that the equations of motion of any mechanical system could be derived from this single principle. Although EULER and LAGRANGE must be credited with formally demonstrating the great generality of the principle of Least Action, it should be recalled that the principle was originally proposed by MAUPERTUIS (1744, 5) the idea having come to him as an extension of FERMAT'S Principle of Least Time (1662), the principle by which FERMAT had attempted to unify optics. LAGRANGE outlined his theory in his "Mécanique Analytique" (1788) which was the first great treastise on analytical mechanics: even today it stands as a landmark in the development of the theory of motion.

About fifty years later, HAMILTON (1834, 6; 1835, 7) made two important contributions to analytical mechanics. First, he showed that LAGRANGE'S equations of motion could be transformed into a very simple symmetrical form—now called the canonical form of the equations of motion. Second, he introduced a very remarkable function which he called the Principal Function by which he completely characterised the motion of any mechanical system. The Principal Function was the natural extension in analytical mechanics of his Characteristic Function, the function by means of which he had earlier unified mathematical optics (1824, 8). Both the Principal Function and the Characteristic Function define a certain wavefront associated with the motion, whose importance will become apparent later.

HAMILTON furthermore showed that the two apparently distinct Sciences of Mechanics and Optics could be unified by the same central principle—the Principle of Stationary Action. Being a mathematician, as well as a physicist, he was impressed by the extraordinary generality of his theory: he felt that these general principles of motion would someday be applicable in other sciences. Thus he observed that: "...the idea, which has thus been exemplified in the sciences of optics and dynamics, seems not confined to these two sciences, but capable of other applications..." (1834, 6). The General Economic Theory of Motion that is here proposed, will be seen to represent a natural extension of the beautiful ideas of EULER, LAGRANGE, and HAMILTON to economic analysis.

3. Lest it should seem strange to allow ourselves to be guided by general principles that underlie a subject so apparently different from economic theory, we may do well to recall at this stage, that the basic Principle of Statics in economic theory is exactly the same as the basic Principle of Statics in analytical mechanics. This was recognised by JEVONS (1871, 9): "The Theory of Economy... presents a close analogy to the science of Statical Mechanics, and the Laws of Exchange are found to resemble the Laws of Equilibrium of a lever as determined by the principle of virtual velocities." And later: "It is curious, moreover, that when we take the theory of the lever treated according to the principle of virtual velocities, we get equations exactly similar in form to those of the theory of value..." Similarily, FISHER (1892, 10) asserts: "The principle underlying the equilibrium of a pendulum or any mechanical equilibrium...is: that configuration will be assumed which will minimise the potential. So also the supreme principle in economic equilibrium is: that arrangement will be assumed which will maximise utility." Since that time it has become fully recognised that the position of rest of an economic system with an objective or utility function, and the position of rest of a

mechanical system with a potential function, both subject to constraints, are characterised by the same property: that the objective function or potential function attain a stationary value, subject to the constraints. The solution to this problem was introduced by LAGRANGE in 1788 in the first part of his "Mécanique Analytique". Thus if

$$U(x_1,\dots,x_n)$$

denotes the objective function or potential function, and

$$\begin{aligned} g^1(x_1,\dots,x_n) &= 0 \\ &\vdots \\ g^k(x_1,\dots,x_n) &= 0 \end{aligned}$$

denote the constraints, then the function

$$U(x_1,\dots,x_n) + \sum_{i=1}^{k} y_i g^i(x_1,\dots,x_n)$$

where $y_1,\dots,y_k$ are the Lagrange Multipliers, attains a stationary value at the position of rest. Thus if $\bar{x} = (\bar{x}_1,\dots,\bar{x}_n)$ denotes the position of rest, then

$$U_x(\bar{x}) + \sum_{i=1}^{k} y_i g^i_x(\bar{x}) = 0$$

where $U_x(\bar{x}) = \left(\frac{\partial U(\bar{x})}{\partial x_1},\dots,\frac{\partial U(\bar{x})}{\partial x_n}\right)$ denotes the gradient of $U(\bar{x})$; this is NEWTON'S characterisation of the position of rest in terms of the forces acting on the system: the position of rest is such that the sum of the forces acting on the system is zero. In passing, we might note that while in mechanics forces were introduced before the potential function, in economic theory marginal utilities were introduced after the utility function. Recognising therefore that the basic Principle of Statics is the same in economics as in mechanics, it may seem less strange to inquire whether the principles of Motion might not be similar in the two subjects. Before proceeding to such an inquiry, however, let us consider the origin of the basic concepts that we will be using, in economic analysis.

4. The idea of basing a general economic theory of motion on a single principle, analogous to the central principle of motion in mechanics seems to have been first proposed by EDGEWORTH (1881, 11): "Energy may be regarded as the central idea of Mathematical Physics; maximum energy the object of the principal investigations of that science. By aid of this conception we reduce into scientific order physical phenomena, the complexity of which may be compared with the complexity which appears so formidable in Social Science .... Now this accumulation (or time-integral) of energy which thus becomes the principal object of the physical investigation is analogous to that accumulation of pleasure which is constituted by bringing together in prospect the pleasure existing at each instant of time, the end of rational action, whether self-interested or benevolent. The central conception of Dynamics and ...in general of Mathematical Physics is...identical with the central conception of Ethics (Economics)..." And later: "'Mécanique Sociale' may one day take her place along with 'Mécanique Celeste', throned each upon...one maximum principle, the supreme pinnacle of moral as of physical science." EDGEWORTH however does not develop this idea any further. Although HOTELLING (1925, 12) refers to the "dawning economic theory (of motion) based on considerations of maximum and

minimum which bears to the older theories the relations which Hamiltonian dynamics and the thermo-dynamics of entropy bear to their predecessors", it would seem that the idea does not reappear in the economic literature until some sixty years later, when AMOROSO (1940, 13), after noting that the equations of motion in mechanics are derived from the Principle of Least Action, observes that: "The parallel principle of Economics is...the principle of maximum satisfaction, which is the basis of all economic activity, and from the beginning it was postulated as the foundation of the doctrine. In this article it constitutes the point of departure, from which the equations of economic movement (analogous to LAGRANGE'S equations) are derived..." The basic principle is stated by him as follows: "...from among all lines that in the same interval of time radiate from a given configuration to reach another given configuration, the individual chooses the one which leads to the maximum value." AMOROSO then proceeds to use this principle to derive equations of motion for the process of production: although stated in a fairly general manner, the principle is not, however, used as a point of departure for a general theory of motion. Although therefore, as AMOROSO indicates, the principle that the motion of an economic system is directed in such a way as to attain maximum benefit to the party concerned, has long been tacitly assumed as the real basis for motion in economic theory it would appear that the principle has never been explicitly used as a point of departure for a general economic theory of motion.

5. It is perhaps appropriate at this stage to recall briefly two important contributions to the economic theory of motion which are very distinct from the present analysis. SAMUELSON in (1941, 14;

1942, 15) and in Chapter XI of the "Foundations of Economic Analysis" (1947, 16) examined certain general stability properties of dynamical economic systems and their relation to static systems. HICKS in Parts III and IV of "Value and Capital" (1938, 17) introduced a theoretical framework for analysing the motion of an economic system under uncertainty. His analysis yielded considerable economic insight into the influence of uncertainty and expectations on the motion of an economic system, however, as HICKS himself discovered, this method of analysis is not easily transformed into a rigorous theory of motion. Even though it is not rigorous however, HICKS' theory has the distinction of being the first formulation of a general stochastic economic theory of motion. The present approach is less general than that of HICKS in that it does not attempt to incorporate the influence of uncertainty on the trajectory of motion. As we have already remarked, however, the approach in these two contributions is quite different from that adopted here.

6. While for the most part the classical economists were more concerned with statics than dynamics, MALTHUS, RAE, and J. S. MILL in particular, (1836, 18; 1834, 19; 1848, 20) made a major contribution to the theory of motion by their careful analysis of one of the great economic problems of motion—the combined effect of population growth, capital accumulation and technological progress on the welfare of a community; their contribution was subsequently extended in particular by BÖHM-BAWERK (1889, 21) and somewhat later by WICKSELL (1901, 22); yet, with the exception of WICKSELL, the analysis in each case was quite unrigorous; almost one hundred years passed from the time of the earlier investigations before the first general and rigorous analysis of the problem appeared. In-

deed RAMSEY'S justly celebrated paper (1928, 23) may be considered as the first accurate mathematical solution of a non-trivial problem of motion, to be published in economic theory, in which the differential equations of motion are explicitly derived. RAMSEY'S contribution gave rise some thirty years later to a whole literature (24-33) on the problem of optimal economic growth beginning with SAMUELSON and SOLOW'S 1956 paper. And now as a result the problem of optimal economic growth stands as the most thoroughly analysed problem of motion in economic theory.

RAMSEY'S analysis is very different from the Keynesian income analysis of the period 1930-1950: this analysis is often supposed to have been an important contribution to the theory of motion; in some ways this is true, but in one significant way it is not. For these income generating systems are not genuine dynamical systems, in so far as the equations of motion are not derived from any central principle. In this respect the Keynesian income system can not be considered as an important contribution to the theory of motion.

7. RAMSEY'S analysis was essentially a pure allocation solution of the problem of motion. However, as we will see later every pure allocation solution may be transformed into a dual valuation-allocation solution. Each problem of motion will have valuation as well as allocation equations. The basic ideas that underlie the valuation equations have a history stretching back many centuries.

The problem of valuation over time had its most direct origin in the practice of borrowing and lending money. Even in primitive societies it was recognised that money has a time value. The concept of present value is implicit in the practice of borrowing and lending. As more complex forms of borrowing and lending emerged, the problem soon presented itself of translating the value of a stream of income or payments into the present. Thus already by the eighteenth century bankers had developed the basic expression for the present value of bonds and annuities (1915, 34). The constant financial practice of reducing all valuations of monetary assets over time to present values hardly had any influence on the economic theory of valuation over time: the economist largely dismissed the concept of present value from his theory of valuation. And then came IRVING FISHER. I. FISHER appears to have been the first economist to propose that <u>the concept of present value should form the basis for the whole theory of valuation over time</u>. "The principle of present worth is of fundamental importance in the theory of value...The...principle which applies here is that the value of capital at any instant is derived from the value of the future income which that capital is expected to yield...the value of capital is the discounted value of its expected income." (1906, 35). HOTELLING'S paper (1925, 12) was one of the earlier mathematical formulations of the same theory: HOTELLING'S analysis may be regarded as a general mathematical theory of valuation over time. The earliest mathematical formulation of FISHER'S theory, in the economic literature, was SAMUELSON'S very important paper (1937, 36): here for the first time the basic differential equation of valuation is shown to be a statement in <u>differential form</u> of the <u>integral Principle of Present Value</u>, which is made the basis of valuation. Much of our analysis may be regarded as a generalisation of the concepts introduced in this fundamental paper.

8. The concept of <u>duality</u> or symmetry between valuation and al-

location is perhaps the single most important and most fruitful concept that mathematical economics has introduced into the formal body of economic theory. It expresses in a new and precise form what every economist has recognised for a long time: that resource allocation and pricing are somehow part of the same problem. The concept was first explicitly introduced by VON NEUMANN (1932) in his remarkable paper on an expanding economy (1938, 37). The concept subsequently reappeared in the highly original analysis of efficient allocation by the Russian mathematician KANTOROVICH (1939, 38), an analysis which contained the first formal statement of linear programming, as well as the first economic interpretation of the constants associated with an optimal allocation, as the valuations imputed to the scarce resources. KANTOROVICH'S paper however remained unknown to the West for almost twenty years, during which time linear programming was rediscovered in the West, among others by DANTZIG (1949, 39). This development inspired KOOPMANS (1951, 40), to reformulate the problem of efficient allocation in the framework of a linear model of production, and to demonstrate rigorously the equivalence between pricing and the efficient allocation of resources. KOOPMANS' fundamental paper thus represents essentially a rediscovery of the earlier results of KANTOROVICH.

These developments led economists to discover the importance for economic theory of a collection of ideas whose origin may be traced to the beautiful researches of MINKOWSKI at the turn of the century (1911, 41). For it became apparent that duality arises essentially from the property of convexity: duality is generated by the separation of convex sets by hyperplanes the coefficients of which denote the imputed prices. This discovery led ARROW

(1950, 42) and DEBREU (1951, 43) to a fundamental reformulation of classical welfare economics: for by extending KOOPMANS' earlier result on duality in efficient production, to the duality between pricing and PARETO optimal allocation of resources in both distribution and production, they were able to demonstrate rigorously the validity of the intuitions underlying the earlier investigations of WALRAS (1874, 44) and PARETO (1897, 45) on the optimality of a competitive price mechanism. Furthermore, while DEBREU showed that the duality theorem could be extended in a straightforward manner to allocation over a finite number of time periods, MALINVAUD, in a beautiful paper (1953, 46) extended the theorem to allocation over an infinite time horizon.

These formal extensions of the duality theorem to allocation over time did not however reveal the essentially <u>recursive</u> nature of production and consumption in an economic system. For commodities may require several time periods to be produced, just as they may yield their benefits in a flow which extends over several periods of time: in such cases valuations must be in the sense of IRVING FISHER, present valuations of future benefits or costs. The first extension of the duality theorem to optimal allocation over time, with valuations representing present valuations of future benefits, occurred in the context of capital theory and not in the context of the classical theory of value. In two fundamental contributions, UZAWA (1964, 26) and SCRINIVASAN (1964, 27) demonstrated, in the framework of a two-sector economy, that associated with each optimal trajectory is an efficient price, which represents the present value of the future stream of marginal benefits generated by an additional unit of investment. This duality was subsequently explored more fully, primarily under the guidance of PONTRYAGIN'S theory of optimal control (1962, 47), in particular in the contributions of CASS (1965, 29; 1966, 30), SHELL (1967, 48) and RYDER (1967, 49). Chapter IV of the present inquiry extends these earlier investigations by demonstrating how the duality between valuation and allocation are embedded in a general economic theory of motion: just as in the earlier static duality theory, convexity and the basic theorems on the separation of convex sets by hyperplanes, are the fundamental properties that generate the duality.

Duality of a rather different kind appeared in the previously mentioned paper by SAMUELSON and SOLOW (1956, 24). But this is the duality of analytical mechanics. After making a significant generalisation of the Ramsey problem, they paused to "review some of the fashionable Hamiltonian formalisms that are often used in the physics and mathematics literature..." This is the first instance in the economic literature where the Hamiltonian theory of analytical mechanics is formally outlined: this is an important step. It should be noted, however, that SAMUELSON and SOLOW appeared to be more concerned with the computational rather than the theoretical aspects of the Hamiltonian formalism, and it is only at the end of the paper that they observed that "it would also be possible to give price interpretations to the Hamiltonian momenta." Furthermore, SAMUELSON and SOLOW did not attempt to develop the theory in the manner that is most suitable for economic analysis, for certain of the functions used in the Hamiltonian formulation of analytical mechanics are not quite appropriate for the Hamiltonian formulation of economic theory; the reasons for this will become apparent in Chapter IV.

9. In seeking for a unified theory of motion in economics, the principal source of inspiration in the present inquiry has been the wonderful theory of motion of material bodies developed during the eighteenth and nineteenth centuries, and known since the work of LAGRANGE (1788, 4) as analytical mechanics. In this respect the recent theory of dynamic programming of BELLMAN (1957, 50), and the theory of optimal control of PONTRYAGIN (1962, 47) do not represent a fundamentally new approach, for these are essentially theories that lie within the mainstream of ideas generated by analytical mechanics. Thus the function which satisfies BELLMAN'S main equation is the same as the Characteristic Function introduced more than a century earlier by HAMILTON, while PONTRYAGIN'S approach leads essentially to HAMILTON'S canonical equations of motion. Both contributions however lead to significant new insights into the theory of motion: thus PONTRYAGIN'S maximum principle leads to a reinterpretation of the transformation of the equations of motion into dual form that has an important economic interpretation, while BELLMAN'S derivation of the fundamental equation gives considerable economic insight into the process involved in transforming the problem of motion into dual form. In this respect the highly original theory of differential or dynamical games developed by ISAACS (1965, 51) leads to insights essentially similar to those obtained from BELLMAN'S theory of dynamic programming.

10. It is on this rich heritage of ideas, part from the mathematical physics of the eighteenth and nineteenth centuries, and part from the twentieth century science of mathematical economics, that the present inquiry is founded.

* * * * *

Chapter II

On the Objective Function of an Economic System

11. Current economic theory is subdivided into a number of different branches, such as the microeconomic theory of value and prices, the macroeconomic theory of income determination, the theories of general equilibrium, welfare economics, international trade, public finance, and monetary theory. Subdivision of subject matter is natural in any science; as the science progresses and increases the breadth of its subject matter the subdivisions adjust to accomodate the increase in understanding: economics has been no exception to this process.

This subdivision of economic theory has enabled economists to obtain a deeper understanding within each of the separate branches, and hence a broader understanding of the whole field. But if this process of subdivision is also accompanied by a subdivision of methods of analysis, then, while the breadth of the field expands faster, the theory takes on an air of unreal complexity, as each branch proceeds to generate its own particular methods.

Although the mathematical analysis of economic problems of motion is of relatively recent origin, as indicated earlier, such a subdivision of methods has already been exemplified in the different dynamical methods of analysis that have been developed within separate branches of economic theory. In many cases, whenever a dynamical problem arises, ad hoc methods are developed to solve the particular problem at hand.

This emergence of different ad hoc methods in the economic theory of motion bears a striking similarity to the early development of the theory of motion in mechanics about two hundred years ago. Then too, particular methods were developed for the solution of each particular problem. The solution of dynamical problems became a veritable exercise in ingenuity. As LAGRANGE observed (1788, 4): "... il fallait toujours une addresse particulière pour démêler dans chaque problème toutes les forces auxquelles il était nécessaire d'avoir égard, ce qui rendait ces problèmes piquants et propres à exciter l'émulation." Then in 1743 D'ALEMBERT (1743, 52) proposed a direct and general method for solving all problems of motion in mechanics—a method which was considerably extended by LAGRANGE. The outcome was one general theory of motion applicable to all problems of motion in mechanics.

It is the purpose of the analysis that follows to indicate, that <u>provided certain conditions are satisfied, a general method also exists in economics, for the solution of problems of motion</u>, the adoption of which renders unnecessary the development of different ad hoc methods. In this way, as far as concerns dynamical problems in economic theory, subdivision of subject matter can proceed without an associated subdivision in the methods of analysis.

12. A distinguishing feature of all economic problems of motion is the presence of <u>uncertainty</u>. It seems probable that there is no dynamical economic problem which is not in some degree influenced by uncertainty. For in economic theory, motion depends principally upon what the <u>future</u> has in store, and in this world, the future is

always shrouded with some measure of uncertainty.

The element of uncertainty is not however of equal importance in all dynamical problems. Two classes of problems may be singled out: first those in which uncertainty is an integral part of the structure, and for which the uncertainty must be explicitly embodied in the theory of motion; second those in which the structure of the motion is not fundamentally altered if the uncertainty is replaced by the ideal conditions of certainty. The former require a general stochastic theory of motion, which economic theory does not yet possess. The latter may be handled by the present theory; for the theory which is here proposed is a theory of motion under the ideal conditions of certainty. This leads us to the <u>first condition</u> that needs to be satisfied in order that the present theory may be applied to a dynamical problem: <u>it must be possible to eliminate uncertainty from the problem without fundamentally altering the nature of the problem, thereby analysing the motion under conditions of certainty</u>.

By way of illustration, consider a few examples from these two classes of problems. As an example from the first category consider the problem of diversification over time. A typical firm does not normally confine its source of revenue to a single product, nor does a typical investor place all his funds in a single type of security. In practice the firm diversifies the array of its products, relying on several products for its revenue, just as the investor diversifies his portfolio by investing his funds in several different securities. Under conditions of certainty the firm will normally be led to produce only a single product, the one that yields the maximum present profit, while the investor will choose to place all his funds in the security with the maximum present value. The phenomenon of diversification arises because the products yield their revenue with different degrees of certainty, just as the securities yield their returns with varying degrees of certainty. Diversification has its origin in uncertainty; a satisfactory explanation for these types of dynamical diversification must therefore be sought in a stochastic theory of motion.

As an example from the second category consider the problem of the motion of a corporation, which will be analysed more fully later. It might seem strange to consider the motion of a corporation under ideal conditions since so much of the structure of a corporation is strongly influenced by uncertainty; indeed the corporation originally emerged as an organisation which could more effectively carry uncertainty than its predecessors, the private firm or the partnership, by limiting liability and dividing ownership among many shareholders. In terms of the theory of the corporation to be developed later, a basic characteristic of the corporation is the proportion of debt in its capital structure. Now it is largely uncertainty over future expected earnings that limits the extent to which a corporation can borrow: if there were no uncertainty in the future earnings profile of the corporation there might be no limit to the extent to which a corporation could borrow, and the analysis of the growth of the corporation would get out of hand. This particular influence of uncertainty, however, can be avoided by supposing from the start that the corporation always carries a certain proportion of debt in its capital structure. In this way, this particular effect of uncertainty—limiting the amount of debt in the capital structure—can be incorporated in the problem without introducing uncertainty explicitly.

Similarly the rate of return which shareholders expect on their stock exceeds the rate of return that bondholders earn on the debt of the corporation. This is because the flow of common stock dividends is less certain than the flow of interest payments on the debt. Under ideal conditions the rates of return on the two assets should be the same; but if no distinction is made between the two rates then the distinction between debt and equity vanishes. It is therefore supposed that for reasons external to the corporation, the rate of return required on the debt is less than that required on the equity. In this way once again, the effect of uncertainty—to create different rates of return on debt and equity—can be incorporated in the problem without explicitly introducing uncertainty.

As a second example, consider the great classical problem of motion, first formalised by RAMSEY, the motion of an aggregate economy under the influence of population growth, capital accumulation, technological change, and the preferences of consumers: this problem will also be considered more fully later. Once again there will clearly be considerable uncertainty present in the problem. It is not possible to know exactly, in advance, for an aggregate economy, how the technology will change, how the preferences of the consumers will alter or in what way the labor force will grow in the future. Considerable insight into the nature of the motion can however be obtained by supposing each of these factors to be known with certainty. The analysis under ideal conditions can serve as a first approximation to what may be expected under a more general stochastic analysis of the motion.

While a theory of motion under ideal conditions is only preliminary to a more general stochastic theory of motion, the analysis of motion under ideal conditions is not for this reason fruitless: even under these primitive conditions certain basic properties of the motion are already revealed.

The requirement of the ideal conditions of certainty is not new in economics. Much of classical economic theory is based on the assumption of perfect competition, which except in more general analyses such as that of ARROW (1953, 53) or DEBREU (1959, 54), is usually used to imply perfect knowledge of market conditions on the part of all consumers and producers; since this is equivalent to requiring the ideal conditions of certainty, it is apparent, that in this respect, the present theory follows the tradition of classical economic theory.

13. In order that a given dynamical problem be susceptible of mathematical treatment, it must be possible to associate with the problem certain measureable quantities, $x_1, \ldots x_n$, in such a way that these variables can be used as the basis for a completely mathematical interpretation of the problem. The dynamical problem must be sufficiently well-defined to be considered as a system, whose state is defined by the variables $x_1, \ldots, x_n$. This system is called the economic system associated with the dynamical problem. In this way, the analysis of a dynamical problem in economics is reduced to the study of an associated economic system, the economic system being the mathematical way of characterising the problem. This concept of an economic system corresponds to the concept of a mechanical system employed in analytical mechanics and should not be confused with the conventional meaning of an 'economic system',

namely a whole economy considered as an entity, although such an economy is clearly an example of an economic system.

14. Just as in the static theory the solution of a problem reduces to a determination of the position of rest of the economic system, so in the dynamical theory the solution of a problem reduces to a determination of the trajectory of the economic system over a certain interval of time. Such solutions may be determined in a relatively simple manner provided the position of rest, or the trajectory of motion are the result of a consistent set of motives that guide the economic system. Thus in the classical theory of consumer behaviour, for example, in order to explain the bundle of commodities $(x_1,\ldots,x_n)$ purchased by a consumer, when the prices of the goods $(p_1,\ldots,p_n)$ and his income (m), are known, it is supposed first that the consumer has a certain ordinal preference function $U(x)$ defined on the space of commodities, which reveals his preference between commodities, and second that the consumer acts rationally according to this preference function, seeking the most preferred commodity bundle subject to his income constraint. The system is thus characterised by a preference function $U(x)$, which indicates the motives driving the consumer and a constraint $\sum_{i=1}^{n} p_i x_i - m \leq 0$, which denotes the restraints imposed by his limited income. Now as is well-known such a preference relation can be shown to exist under fairly general conditions [HOUTHAKKER (1950, 55)]. More generally, the position of rest of a static economic system is determined by associating with the system an objective function $U(x)$, which denotes the motives that guide the system and certain constraints $g^i(x) \geq 0$, $i = 1,\ldots,k < n$, which indicate

the manner in which the motives of the system are restrained.

While in the static case the benefits that accrue to the economic system depend only on its position, in the dynamical case, the benefits that accrue to the system at any instant, may depend not only on its position $x = (x_1, \dots, x_n)$, but also on its direction of motion $\dot{x} = \frac{dx}{dt}$, as well as on the time t. Thus the objective function at any instant may be written as

$$U(t,x,\dot{x}) \qquad (1)$$

while the constraints, which may also depend on the position, direction of motion and the time, may be written as

$$g^i(t,x,\dot{x}) \geq 0 \qquad i = 1,\dots,k < n \qquad (2)$$

The objective function denotes the benefits or utility accruing to the system at each instant. The objective function and the constraints are the basic functions associated with the economic system from which all other functions that characterise the motion are derived. Note that in certain economic systems the objective function is separable in the form

$$U(t,x,\dot{x}) = f(t).u(x,\dot{x}) = e^{-\int_0^t \delta(z)dz} u(x,\dot{x}) \qquad (1)'$$

where $\delta(t)$ denotes a variable discount rate.

Since the present theory requires that such an objective function and constraints may be associated with the economic system, the second condition for the applicability of the theory may be stated as follows: it must be possible to transform the analysis of the dynamical problem into the analysis of an economic system, which may be characterised by an objective function $U(t,x,\dot{x})$

and constraints $g^i(t,x,\dot{x}) \geq 0$, $i = 1,\ldots,k < n$.

15. By the method first demonstrated by VALENTINE (1937, 56), the inequality constraints (2), may be converted into the equality constraints

$$g^i(t,x,\dot{x}) - (\dot{\xi}_i)^2 = 0 \ , \quad i = 1,\ldots,k$$

by introducing the new variables $\dot{\xi}_i$. By virtue of the elegant method, first introduced by EULER (1736, 57) and subsequently extended by LAGRANGE (1788, 4), a system with equality constraints can be transformed into a system free of such constraints. Thus by introducing the variables $y_1,\ldots,y_k$, called the Lagrange multipliers associated with the constraints $g^1,\ldots,g^k$, and forming the new objective function

$$\bar{U}(t,x,\dot{x},\dot{\xi},y) = U(t,x,\dot{x}) + \sum_{i=1}^{k} y_i [g^i(t,x,\dot{x}) - (\dot{\xi}_i)^2] \qquad (3)$$

it can be shown that the system with objective function and constraints (1) and (2) is equivalent to the system free of constraints with objective function (3). In view of this property the subsequent analysis may be confined to economic systems free of constraints. For notational simplicity, we may, without loss of generality, write the objective function as $U(t,x,\dot{x})$ rather than $\bar{U}(t,x,\dot{x},\dot{\xi},y)$.

Since the economic system may now be characterised by a single objective function, the second condition may be referred to as the objective function condition. If the first condition is called the certainty condition, then we obtain the following characterisation of the conditions under which the present theory may be applied:

a dynamical problem may be solved by the methods of the present theory provided it satisfies both the certainty and the objective function conditions.

Just as in static economic systems, so in the present dynamical systems, the Lagrange multipliers $y_1, \ldots, y_k$, denote imputed values or valuations induced by the constraints. The constraints in an economic system thus introduce a whole system of valuation. In the present inquiry, however, we will not attempt an analysis of this constraint-induced valuation, concentrating rather on yet another system of valuation induced by the problem of motion, which will be introduced in Chapter IV.

16. The theory may be illustrated by two simple examples, both of which may be formalised so that these two conditions are satisfied.

Consider first the motion of a corporation mentioned earlier. The dynamical problem involves analysing the way a corporation expands over time, how its growth is influenced in particular by dividend and debt policy and how these factors influence the price of its common stock. The problem may be formalised as follows.

Let $k(t)$ denote the permanent capital on the equity side of the balance sheet of the corporation: $k(t)$, which will be used to indicate the size of the corporation, consists of all long-term liabilities or debt, $b(t)$, with maturity date in excess of say, one year, and common equity, namely capital stock, capital surplus and earned surplus. The profits of the corporation consist of trading revenue less depreciation, production costs and profits taxes. The profits that the corporation earns depend upon the way the permanent capital is used, as reflected on the asset side of the balance sheet, the nature of the corporation itself and the industry in which it is operating. The chief characteristic of profits, however, is that they are always to some extent uncertain. The uncertainty is strongly influenced by the industry in which the corporation is operating; thus in the utility industry the uncertainty is considerably smaller than in most branches of manufacturing. Uncertainty apart, the profits will be a function of the size of the corporation. The managers of the corporation will know, at least approximately, the form of this profit function $f(k)$. To apply the present theory, however, we must suppose that the managers know exactly and correctly the form of the function $f(k)$. Figures 1 and 2 illustrate two typical profit functions. In the first case, profits increase rapidly at first, after which further

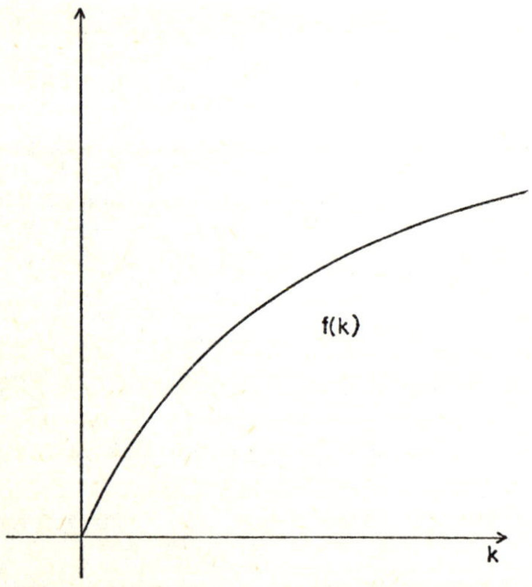

Figure 1. First example of profit function

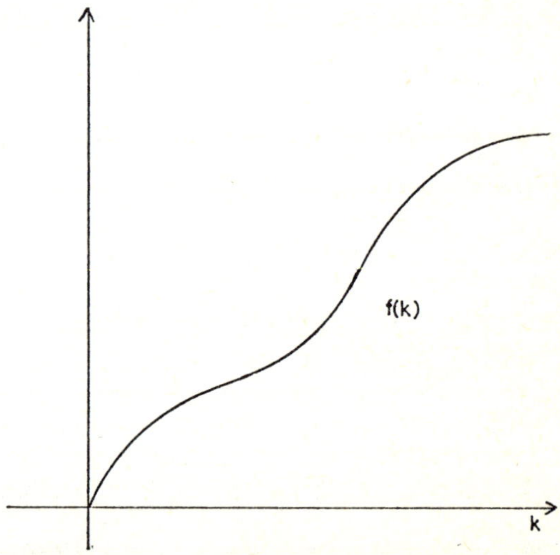

Figure 2. Second example of profit function

size increases lead to smaller and smaller profit increases. The second case is similar to the first for an initial phase, after which profits again rise rapidly with increases in size. In the first case the corporation faces saturation: in the second case after facing an initial phase of saturation, by further expanding its permanent capital, for example by moving into a larger scale of production, the corporation moves to a second phase of rapidly increasing profits.

The corporation can expand, or increase its permanent capital in four principal ways: by borrowing, by retaining some of its profits, by issuing equity and by merger. In order to simplify the analysis, the present formulation will exclude growth by merger and growth by the issue of new equity. For reasons indicated earlier, we suppose that the corporation always carries a certain fixed proportion, β , of debt in its capital structure, where $0 \leq \beta < 1$. Thus provided that it is to the benefit of the corporation to grow βk of the permanent capital will always be debt, and its borrowing will always be $\dot{b}(t) = \beta\dot{k}(t)$. If $\alpha > 0$ denotes the constant interest rate payable on the debt, then interest payment on the debt is $\alpha\beta k(t)$. Profits are either distributed in dividends $d(t)$, or retained, in which case they are used to increase permanent capital $\dot{k}(t)$. Thus at each instant the following accounting identity must be satisfied: profits plus borrowing must equal interest payment on the debt, plus dividend payments, plus increase in permanent capital, which implies that

$$f(k) + \beta\dot{k} = \alpha\beta k + d + \dot{k} \qquad (4)$$

While the shareholders are the effective owners of the corporation, it is the managers that control its operations over time. In most cases the managers try to control the operations of the corporation in the way that will be most beneficial to the shareholders. The two benefits that shareholders obtain are dividends and increases in the market price of the stock. To simplify the analysis we will suppose that shareholders are indifferent between these two ways of receiving benefits: the shareholders are thus indifferent between a given increase in the price of the stock, or the equivalent amount of dividends. Now under the present ideal conditions, the price of the stock depends on the dividends, for the price is equal to the present value of the future stream of dividends. The objective function of the corporation may thus be formulated in terms of the stream of dividends. Although the managers cannot know with certainty, the benefits that accrue to shareholders from a given amount of dividends, they form an estimate of these benefits, $u(d)$, bearing in mind that shareholders always prefer more dividends, but at a decreasing rate,

$$u'(d) > 0 \ , \ u''(d) < 0 \ , \ d \geq 0 \ ,$$

and that shareholders prefer a given amount of dividends this year, to the same amount of dividends next year. We will suppose that this rate of time-preference $\delta(t)$ is positive and constant. The objective function thus becomes

$$u(d(t)).e^{-\delta t}$$

As explained earlier, the return δ on equity must exceed the interest rate α on the debt, in order to induce a capital structure which contains both debt and equity: we thus suppose that the corporation can borrow from a special loan market on which the interest rate α is lower than the time-preference δ of the shareholders. Since (4) implies

$$d = f(k) - \alpha\beta k - \dot{k}(1-\beta) \tag{5}$$

the objective function is

$$u(f(k) - \alpha\beta k - \dot{k}(1-\beta)).e^{-\delta t} \tag{6}$$

which is a special case of the separable objective function (1)'. The objective function (6) is the basic function associated with the corporation, in terms of which the whole subsequent analysis will be developed, there being no need, in this case to introduce any constraints.

17. The second example, which was also mentioned earlier, is the classical problem of the motion of an aggregate economy as formalised by RAMSEY.

The dynamical problem involves analysing the efficient way in which an aggregate economy should expand over time and how this expansion is influenced in particular by the distribution of income between consumption and saving. The problem may be formalised as follows.

Consider an aggregate economy whose motion is under the guidance of a Planning Authority. The economy is endowed with a labor force $L(t)$ which the Authorities estimate will continue to expand at a proportional rate $\nu > 0$, and a capital stock $K(t)$ of a single good which is known to depreciate at a proportional rate $\mu > 0$, independent of its use. While the technology cannot be known with certainty, the best estimate is contained in a production function $F(K,L)$ which shows the output of the single good that can be obtained from given combinations of capital and labor. To simplify the analysis it is assumed that $F(K,L)$ is homogeneous of degree one in K and L , and that there is no technical progress.

The output of the good is allocated either towards consumption by the individuals in the community, or towards increasing or maintaining the capital stock. This implies, introducing the per-capita variables, $k(t) = \frac{K(t)}{L(t)}$, $c(t) = \frac{C(t)}{L(t)}$, and the function $f(k) = F(k,1)$,

$$f(k) = c + \dot{k} + \lambda k \tag{7}$$

where $\lambda = \nu + \mu$, and λk is the investment-per-worker required to maintain the capital-per-worker at its current level. In terms of $f(k)$, the technology is such that the marginal product of capital-per-worker is always positive and decreasing,

$$f'(k) > 0 \ , \ f''(k) < 0 \ , \ k \geq 0$$

The Planning Authority controls the economy in the manner that is most beneficial to the members of the community. The sole benefit that accrues to an individual in the community comes from consuming the only good that is available. The Planning Authority estimates that the members of the community have approximately identical tastes and that a typical consumer's preference is given by the function

$$u(c(t)).e^{-\rho t}$$

where

$$u'(c) > 0 \ , \ u''(c) < 0 \ , \ c \geq 0 \ ,$$

the marginal utility of consumption is always positive, but decreasing, and where $\rho > 0$ denotes his constant rate of time-preference. The utility accruing to an individual is assumed to depend only on his own consumption. The welfare of the whole community is therefore, at any time,

$$L(t).u(c(t)).e^{-\rho t} = L(0).u(c(t)).e^{-(\rho-\nu)t}$$

and this function is taken as the objective function for the system.

Since (7) implies

$$c = f(k) - \dot{k} - \lambda k$$

the objective function becomes

$$L(0).u(f(k) - \dot{k} - \lambda k).e^{-(\rho-\nu)t}$$

which is again a special case of the separable objective function (1)'.

This very simple example may at first appear to be of limited value in view of the highly aggregated manner in which the problem has been formulated. A generalisation to the case of an m-commodity system reveals, however, that the essential nature of the problem remains unchanged in the m-commodity economy. Although there are a number of ways in which this generalisation may be carried out, the simplest method is the one first introduced by SAMUELSON and SOLOW (1956, 24). In this formulation it is assumed that $\nu = \mu = 0$; the production equation

$$c + \dot{k} = f(k)$$

is replaced by the production frontier

$$c_1 + \dot{k}_1 = f(k_1, \ldots, k_m \; ; \; c_2 + \dot{k}_2, \ldots, c_m + \dot{k}_m)$$

so that the objective function for the system becomes

$$L(0).u(f(k_1,\dots,k_m \; ; \; c_2+\dot{k}_2,\dots,c_m+\dot{k}_m) - \dot{k}_1, c_2,\dots c_m).e^{-\rho t}$$

Further analysis of the problem reveals that $c_1,\dots c_m$ must be eliminated by maximising with respect to these variables, the objective function thus reduces once again to a special case of (1)′.

* * * * *

Chapter III

On the Principle of Stationary Present Value

18. In economic theory there are two main kinds of dynamical problems: in the first, the object is to explain and predict certain observed behavior; these may be called descriptive or predictive problems. In the second, the object is to prescribe what action should be taken in order that a given objective be attained; these may be called prescriptive problems. The present theory is applicable to both types of problems provided that in the case of descriptive problems one further condition is satisfied.

For in order to analyse the descriptive or predictive problems something must be known about the basic motives that underlie the observed behavior: without a knowledge of these basic motives no explanation or prediction is possible. In classical economic theory these motives are summarised in a very simple Axiom. The whole of the classical theory is founded on the axiom of economic behavior known as the Axiom of Rational Choice under certainty. The Axiom may be stated as follows: an economic system guided by rational choice under conditions of certainty acts in the manner that assures maximum benefit. The theory of a perfectly competitive economic system, for example, is founded on this Axiom which in this case implies that consumers are motivated to maximise their utility, while producers are motivated to maximise their profits.

While the Axiom represents a strong simplification of the behavioral motives that underlie economic systems, it still remains the most natural starting point for a theory of motion. In the present theory, therefore, this Axiom will be taken to represent the basic behavioral motive underlying descriptive problems: for these problems, therefore, the Axiom of Rational Choice represents the <u>third condition</u> that needs to be satisfied.

While in descriptive problems the basic behavioral motives are largely hidden, in prescriptive problems the motive of the system is stated explicitly. Provided the problem is appropriately formulated, it is clear that a prescriptive problem immediately satisfies the Axiom of Rational Choice, for in this case also that action is chosen which leads to the maximum benefit. The Axiom of Rational Choice may thus be taken as the basic Axiom on which the present theory is founded.

19. This Axiom leads directly to the basic <u>Principle of Motion</u> if the benefits accruing to an economic system at each instant can be characterised by an objective function

$$U(t,x,\dot{x}).$$

For the total benefit accruing to the system, over a finite or infinite interval of time $[0,T]$,

$$\int_0^T U(t,x,\dot{x}).dt \qquad (1)$$

depends on the position and direction $(x,\dot{x})$ chosen by the system at each instant from time 0 to time T, or what is equivalent, to the <u>trajectory $x(t)$</u> of the system during this time interval. Thus when $U(t,x,\dot{x})$ denotes positive benefits the Axiom of Rational Choice implies that the system will choose the trajectory that makes (1) as large as possible, while in cases where $U(t,x,\dot{x})$ denotes negative benefits or costs, the trajectory will be chosen so as to make (1) as small as possible; similarily in cases of conflict, such as arise in dynamical or differential games, where the system is partly under the control of an agent that wishes to maximise (1) and partly under the control of an agent that wants to minimise (1), the integral will be made a maximum with respect to say $x_1,\ldots,x_i$, and a minimum with respect to $x_{i+1},\ldots,x_n$. Now since in a typical economic system, the integral

$$\int_0^T U(x,\dot{x},t).dt = \int_0^T u(x,\dot{x}).e^{-\int_0^t \delta(z).dz}.dt$$

denotes the present value of the whole stream of benefits or costs accruing during the interval of time $[0,T]$, it seems reasonable to refer to the integral (1) in the more general case also, as the Present Value of the system. The basic Principle of Motion may then be stated as follows:

The Principle of Stationary Present Value.

The trajectory $x(t)$ of an economic system with an objective function, over any interval of time $[0,T]$, is such that the Present Value of the system

$$P(x) = \int_0^T U(x,\dot{x},t),dt \tag{2}$$

attains a stationary value.

20. This Principle leads at once to the equations of motion for the economic system

$$U_{x_i} - \frac{d}{dt}(U_{\dot{x}_i}) = 0 \qquad i = 1,\ldots,n \qquad (3)$$

known as the EULER equations (1736, 57), which must be satisfied if the functional (2) is to attain a stationary value.

In the static case the Principle of Motion coincides with the basic Principle of Statics mentioned in Chapter I. In this case the position of rest x is such that the objective function $U(x)$ attains a stationary value; this leads at once to the basic equations of statics

$$U_{x_i} = 0 \qquad i = 1,\ldots,n$$

which are a special case of (3), and are equivalent to the equations mentioned in Chapter I.

The equations (3) will be called the pure allocation equations of motion. For the equations (3) solve the dynamical problem by treating it as a problem of pure allocation, there being no mention of any process of valuation. In this respect the equations (3) are deficient, for they do not reveal the beautiful and powerful theory of valuation that is contained in a very natural way in the problem of motion: this will become apparent in Chapter IV.

21. Any curve that satisfies the Principle of Stationary Present Value must satisfy the equations of motion: these equations do not however distinguish between the maxima and the minima of the functional (2). In order to make such a distinction, a further condition must be satisfied by the curve, known as the LEGENDRE-CLEBSCH condition (1786, 58; 1858, 59). Thus if the trajectory is to lead to a maximum of the integral (2), then along this curve

$$\sum_{i,j=1}^{n} U_{\dot{x}_i \dot{x}_j} \alpha_i \alpha_j \leq 0 \tag{4}$$

must hold for all real numbers α_i, α_j , which are not all zero, and if the trajectory is to lead to a minimum of (2), then

$$\sum_{i,j=1}^{n} U_{\dot{x}_i \dot{x}_j} \alpha_i \alpha_j \geq 0 \tag{5}$$

must hold for similarily chosen α_i, α_j . In the case of a mini-max value (4) must be satisfied by the maximising variables, and (5) by the minimising variables.

The condition (4) represents the analogue in the theory of motion, of the classical condition of diminishing marginal utility in the theory of the consumer, or diminishing marginal productivity in the theory of the producer. As in the classical theory, (4) can be given a simple geometric interpretation.

For the Legendre condition (4) requires that at each instant t , and for each point $x(t)$ on the trajectory of motion, the function $U(t,x,\dot{x})$ be a concave function in $\dot{x}$: this function, which we may write as $U(\dot{x})$, defines a surface in the $(n+1)$ dimensional space $(U,\dot{x})$, which will be called the indicatrix, since it represents the analogue in the present theory of a similar surface first introduced in the homogeneous theory of the calculus of variations [CARATHEODORY (1904, 60)]. Since $U(\dot{x})$ is a concave function, the indicatrix defines a convex region $\mathcal{J}$, in the $(n+1)$ dimensional space $(U,\dot{x})$, $\mathcal{J} \subset \mathbb{E}^{n+1}$,

$$\mathcal{J} = \{(U,\dot{x}) \mid U \leq U(\dot{x}),\ \dot{x} \in D\}$$

where D denotes the convex domain in E^n on which $U(\dot{x})$ is defined. Note that if $U(\dot{x})$ is a convex function as implied by (5), then the convex region $\mathcal{J}$ is defined by

$$\mathcal{J} = \{(U,\dot{x}) \mid U \geq U(\dot{x}),\ \dot{x} \in D\} .$$

As will be shown in Chapter IV, the indicatrix and the associated convex region $\mathcal{J}$ play a fundamental role in the transition to the dual theory of motion.

An example of the indicatrix and the convex region $\mathcal{J}$, may be obtained by using the second example of the last chapter. Thus if $u(c) = \sqrt{c}$, then the indicatrix becomes

$$U(\dot{k}) = e^{-\delta t}\sqrt{f(k) - \lambda k - \dot{k}}$$

where $\delta = \rho - \nu$. Since $-k(t) \leq \dot{k} \leq f(k) - \lambda k$, D is the interval $[-k(t),\ f(k) - \lambda k]$. This surface and the convex region $\mathcal{J}$ are shown in Figure 7 of Chapter IV.

22. The pure allocation theory of motion developed so far may also be illustrated by the two examples introduced in Chapter II.

Consider first the motion of a corporation. According to the Principle of Stationary Present Value, the trajectory of the corporation over any interval of time $[0,T]$ is such that the Present Value of the system

$$P(k) = \int_0^T u(f(k) - \alpha\beta k - \dot{k}(1-\beta)).e^{-\delta t}.dt$$

attains a maximum. This leads to the equation of motion for the corporation

$$u'[f'(k) - \alpha\beta].e^{-\delta t} + \frac{d}{dt}(u'(1-\beta).e^{-\delta t}) = 0$$

which reduces to the second order differential equation

$$\ddot{k}(1-\beta)^2 u'' - \dot{k}(1-\beta)u''(f'(k) - \alpha\beta) - u'[f'(k) - \alpha\beta - \delta(1-\beta)] = 0 \quad (6)$$

(6) is more simply analysed by reducing it to an equivalent first order system. Thus introducing the new variable $v = \dot{k}$, (6) becomes

$$\dot{k} = v$$
$$\dot{v} = \frac{v(f'(k) - \alpha\beta)}{1 - \beta} + \frac{u'[f'(k) - \alpha\beta - \delta(1-\beta)]}{(1 - \beta)^2 u''} \quad (7)$$

The main properties of the solutions to (7) may be found by employing the geometric method originally introduced by POINCARÉ (1881, 61). In this case the method reduces to finding the critical points $\dot{k} = \dot{v} = 0$, and analysing the nature of the linear approximation to (7) in the neighborhood of each of these critical points.

From (7) it is clear that a critical point requires $v = 0$, and

$$\frac{u'[f'(k) - \alpha\beta - \delta(1-\beta)]}{(1-\beta)^2 u''} = 0$$

Since $u' > 0,\ u'' < 0,\ 0 \le \beta < 1$, this implies that

$$f'(k) - \alpha\beta - \delta(1-\beta) = 0 \quad (8)$$

Consider first the case where the corporation has a profit function of the form shown in Figure 1, Chapter II, satisfying the conditions

$$f(k) \geq 0, \; f'(k) > 0, \; f''(k) < 0, \; k \geq 0 \qquad (9)$$

Provided that

$$f'(0) > \alpha\beta + \delta(1-\beta)$$

the only case of economic interest, then there exists only one $\hat{k} > 0$ satisfying (8), since

$$\alpha\beta + \delta(1-\beta) > 0 \;,$$
$$f'(k) > 0, \; f''(k) < 0, \; k \geq 0 \;.$$

$\hat{k}$ may be found in the manner illustrated in Figure 3.

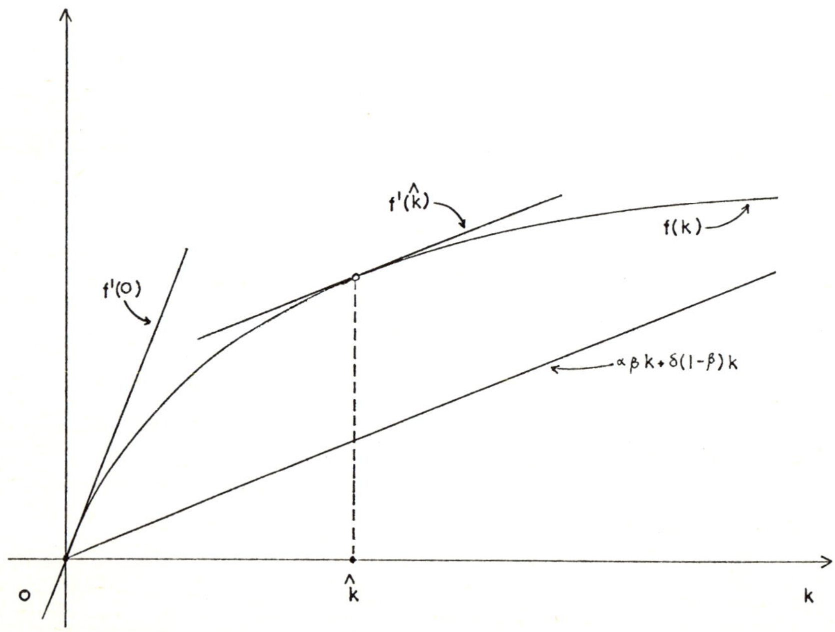

Figure 3. Location of the critical point

The linear approximation to (7) in a neighborhood of the critical point $(\hat{k},0)$ is then

$$\begin{bmatrix} \dot{k} \\ \\ \dot{v} \end{bmatrix} = \begin{bmatrix} 0 & 1 \\ \\ \dfrac{u'f''(\hat{k})}{(1-\beta)^2 u''} & \dfrac{f'(\hat{k}) - \alpha\beta}{1-\beta} \end{bmatrix} \begin{bmatrix} k - \hat{k} \\ \\ v - 0 \end{bmatrix}$$

for which the characteristic equation is

$$r^2 - \delta r - \frac{u'f''(\hat{k})}{(1-\beta)^2 u''} = 0$$

the roots of which are

$$r_1, r_2 = \frac{1}{2}\left[\delta \pm \sqrt{\delta^2 + \frac{4u'f''(\hat{k})}{(1-\beta)^2 u''}}\right] \tag{10}$$

$r_1 > 0$, $r_2 < 0$ since

$$u' > 0 \ , \ u'' < 0 \ ,$$

$$(1-\beta)^2 > 0 \ , \ f''(\hat{k}) < 0 \ .$$

The solution curves thus form a saddle-point in a rectangular neighborhood of the critical point. Since $u(d)$, $d \geq 0$ and $f(k)$, $k \geq 0$ are both concave functions, the neighborhood may be extended to an arbitrarily large neighborhood in the $v - k$ plane, $k \geq 0$, as indicated in Figure 4.

While the basic form of the solution curves is unchanged, the linear approximation becomes increasingly inaccurate as the neighborhood is extended.

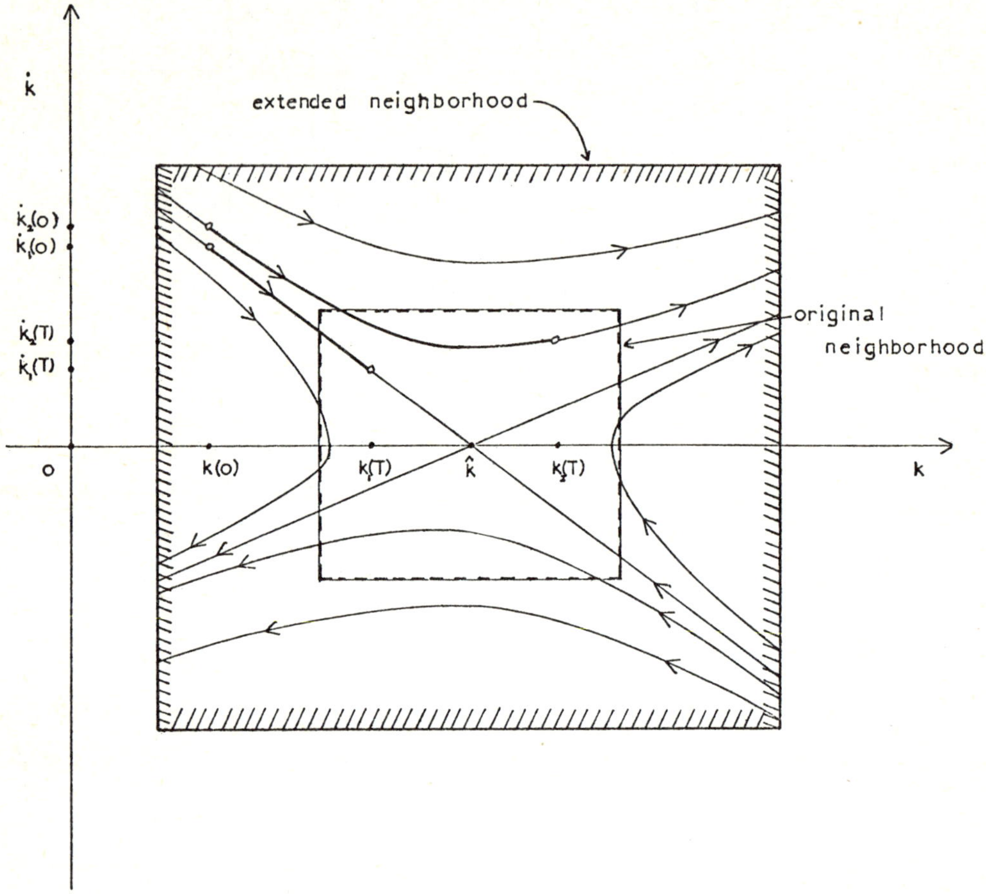

Figure 4. Motion of the corporation

23. Consider the economic interpretation of $\hat{k}$. Just as the interest payment on the debt is $\alpha\beta k$, so the basic dividend that should be earned by the equity, at least in the long-run, is $\delta(1-\beta)k$: if both $\alpha\beta k$ and $\delta(1-\beta)k$ are considered as basic costs of holding the permanent capital, then

$$f(k) - \alpha\beta k - \delta(1-\beta)k$$

may be called <u>net (long-term) profit</u>. The size k^* that makes net (long-term) profit a maximum, may be called the <u>optimal size</u> of the corporation. k^* thus satisfies (8) and $f''(k^*) < 0$, which are exactly the conditions satisfied by $\hat{k}$. Since $\hat{k}$ is unique, $\hat{k} = k^*$. The size $\hat{k}$ is thus the optimal size of the corporation.

Consider the economic interpretation of equation (8), an equation which is very familiar in the theory of the corporation. $f'(k)$ denotes the marginal return on the permanent capital, while $\alpha\beta + \delta(1-\beta)$ denotes the marginal cost of the capital. When $k < k^*$, $f'(k) > \alpha\beta + \delta(1-\beta)$ the marginal return exceeds the marginal cost, and net profit can be increased by increasing the size of the corporation: conversely, when $k > k^*$, the marginal cost exceeds the marginal return and net profit may be increased by decreasing the size. Only when the marginal return is equal to the marginal cost of the capital is net profit at a maximum.

Note the factors that determine the optimal size of the corporation. Given the profit function $f(k)$, the optimal size depends on its <u>capital structure</u> $(\beta, 1-\beta)$, and on the two <u>cost rates</u> (α, δ) , the interest rate on the debt and the time-preference of the shareholders. Since $\alpha < \delta$, increasing the proportion of debt in the capital structure increases the optimal size of the corporation: similarly an increase in the interest rate or time-preference of the shareholders reduces the optimal size.

If the managers could know with certainty the profit function, the preferences of the shareholders, the capital structure and the interest rate on the debt, for the indefinite future, then they could plan the growth of the corporation for all time. But this situation never arises, since the basic information they require can be predicted for only a limited number of years, say T years, into the future. This forces the managers to make periodic plans, the period of the plan being the foreseeable future, T years.

Consider the way in which the managers arrive at such a plan. Suppose $k(0) < k^*$ is the permanent capital inherited from the previous period. $f(k)$, $u(d).e^{-\delta t}$, α and β are known for the next T years and are not expected to change in any definite way thereafter. The corporation may then be one of two principal kinds: in the first the managers act entirely in the interest of the shareholders; in the second, while the managers act in the interest of the shareholders, the growth of the corporation becomes an important objective for its own sake, quite apart from the benefits it may or may not bestow on the shareholders.

In the first case the managers plan the growth of the corporation so that it expands towards the optimal size $k^* = \hat{k}$, along the unique trajectory leading to $(\hat{k},0)$ in Figure 4. Thus given $k(0)$, $\dot{k}_1(0)$, the initial growth rate for this first type of corporation is immediately determined, as are $k_1(t)$ and $\dot{k}_1(t)$ for the rest of the planning period, as indicated in Figure 4. Note that the average growth rate of this first type of corporation during the whole period is

$$\bar{r}_1 = \frac{1}{T}\int_0^T r_1(t).dt$$

where $r_1(t) = \frac{\dot{k}_1(t)}{k_1(t)}$.

The simplest method of formulating the growth of the second kind of corporation is to suppose that in these corporations the managers plan the expansion so as to maximise the welfare of the shareholders, while at the same time maintaining an average growth rate $\bar{r}_2 > \bar{r}_1$. Given $\bar{r}_2$, this implies that

$$k_2(T) = k(0)e^{\bar{r}_2 T}$$

must be attained during the planning period. Since $k(0)$ and $k_2(T)$ are fixed, this at once leads to a unique trajectory in Figure 4, from which $\dot{k}_2(0)$ is determined as well as $k_2(t)$ and $\dot{k}_2(t)$ for the remainder of the period.

Note that in this second case, $\bar{r}_2$ must be chosen so as to be less than the maximum attainable average growth rate $\tilde{r}$, obtained by paying no dividends during the whole period: $\tilde{r}$ is defined by the relation

$$\tilde{k}(T) = k(0)e^{\tilde{r}T}$$

where $\tilde{k}(T)$ denotes the maximum attainable size; $\tilde{k}(T)$ in turn, is defined by the relation

$$T = (1-\beta)\int_{k(0)}^{\tilde{k}(T)} \frac{dk}{f(k) - \alpha\beta k}$$

Certain general characteristics of the growth of a corporation can now be observed, provided the profit function satisfies the conditions (9). A corporation with a given capital structure, faced with a given marginal cost of capital, begins its life with a permanent capital considerably below its optimal level. Initially, the marginal return on capital being much greater than its marginal cost, a large amount of profits are retained, leading to a rapid rate of growth, and correspondingly small dividend payments. As the size of the corporation increases the excess of the marginal return over the marginal cost of capital decreases, and the amount of profits retained decreases, leading to a decreasing rate of growth. The growth of the corporation leads to an increase in its profits and although interest payment on the debt increases, the increase in profits, combined with the decreasing amount of profits retained, leads to an increase in dividends.

The growth of a corporation always slows down as it approaches the optimal size. Provided the corporation is not growth-oriented, it continues to expand until it reaches the optimal size, where it remains indefinitely, paying out in dividends all profits that remain after interest payment on the debt. If the corporation is growth-orientated it expands beyond the optimal size, its rate of growth, and hence the amount of profits retained always being greater than in a corporation acting solely in the interests of the shareholders. The size to which the corporation grows depends on the extent to which the managers value growth in the face of continually declining net profit.

Since the manner in which the managers plan the growth of a corporation determines the dividend payments to the shareholders, a knowledge of this growth plan for the entire future, would enable an accurate valuation to be placed upon the common equity of the corporation; this value $v(t)$, would be the present value of the entire stream of future dividends

$$v(t) = \int_t^\infty (f(k) - \alpha\beta k - \dot{k}(1-\beta)).e^{-\delta(\tau-t)}.d\tau$$

assuming a market discount rate equal to the time-preference of the shareholders. If m is the number of shares originally issued, then the price of a share would be $\frac{v(t)}{m}$. Thus in the ideal case of a corporation which expands to its optimal size, and then remains there indefinitely, it is clear that on reaching the optimal size, the price of a share would be $\frac{f(k^*) - \alpha\beta k^*}{\delta m}$.

The analysis of the growth of a corporation has so far been confined to the case where the profit function satisfies condition (9). Introducing the profit function of Figure 2 in Chapter II shows how radically the motion may be altered by a change in the profit function. In this case provided $f'(0) > \alpha\beta + \delta(1-\beta)$, there are three critical points $(\hat{k}_1,0)$, $(\hat{k}_2,0)$, $(\hat{k}_3,0)$, where $\hat{k}_1 < \hat{k}_2 < \hat{k}_3$, the motion around $(\hat{k}_1,0)$ and $(\hat{k}_3,0)$ being identical with the motion around $(\hat{k},0)$ above. At $\hat{k}_2$ however, condition (8) is satisfied and $f''(\hat{k}_2) > 0$. Net profit thus attains a local minimum at $\hat{k}_2$, while at $\hat{k}_1$ and $\hat{k}_3$ net profit attains a local maximum. $\hat{k}_2$ is thus a very undesirable size for the corporation. Examination of the roots in (10), reveals that the growth rate always increases as it approaches $\hat{k}_2$, the motion around $(\hat{k}_2,0)$ being that of an unstable node or focus, depending on whether the roots are real or imaginary.

24. Secondly, consider the example of the motion of an aggregate economy. According to the Principle of Stationary Present Value, the trajectory of the economy, over any interval of time $[0,T]$, is such that the Present Value of the system

$$P(k) = \int_0^T u(f(k) - \dot{k} - \lambda k).e^{-\delta t}dt$$

where $\delta = \rho - \nu$, attains a maximum. This leads to the equation of motion for the economy

$$u'(f'(k) - \lambda)e^{-\delta t} + \frac{d}{dt}(u'e^{-\delta t}) = 0$$

which reduces to the second order differential equation

$$\ddot{k}u'' - \dot{k}u''(f'(k) - \lambda) - u'[f'(k) - (\lambda+\delta)] = 0$$

which in turn is equivalent to the first order system

$$\dot{k} = v$$

$$\dot{v} = v(f'(k) - \lambda) + \frac{u'}{u''}[f'(k) - (\lambda+\delta)] \qquad (11)$$

the critical points of which occur when $v = 0$ and

$$f'(k) - (\lambda + \delta) = 0 \qquad (12)$$

As before, provided $f'(0) > \lambda + \delta$, there exists only one k^* satisfying (12), since $\mu + \rho > 0$, $f'(k) > 0$, $f''(k) < 0$, $k \geq 0$. The roots characterising the linear approximation to (11) around the critical point $(k^*,0)$ are

$$r_1, r_2 = \frac{1}{2}\left[\delta \pm \sqrt{\delta^2 + \frac{4u'}{u''}\cdot f''(k^*)}\right]$$

where again $r_1 > 0$, $r_2 < 0$ since $u' > 0$, $u'' < 0$, $f''(k^*) < 0$. The solution curves thus form a saddle-point around $(k^*,0)$. This leads at once to the principal characteristic of the motion of the aggregate economy, first pointed out by SAMUELSON (1949, 62) and subsequently referred to as the Turnpike Theorem: provided the period T over which the expansion of the economy is being planned, is sufficiently large, the economy spends an arbitrarily large proportion of the time in a given neighborhood of the critical point $(k^*,0)$. The Turnpike k^*, which is known as the Golden Rule path, represents the optimal amount of capital-per-worker, since at k^* the net marginal product of capital-per-worker $f'(k^*) - \mu$ is equal to the marginal cost of capital-per-worker determined by the rate of time-preference ρ, of the community.

25. It is perhaps appropriate to recall the similarity between the present economic theory of motion and the theory of motion in

analytical mechanics. For just as the basic Principle of Statics is the same in economics as in mechanics, so the basic Principle of Motion is the same in both subjects. The Principle of Motion in analytical mechanics, which corresponds to the Principle of Stationary Present Value in economics, is the Principle of Least Action or the Principle of Stationary Action as it is more accurately called. This Principle leads to a functional similar to (2), the objective function being replaced by the Lagrangean $L(t,x,\dot{x})$ of the mechanical system; the equations of motion (3) are then known as Lagrange's equations of motion for the mechanical system (1788, 4).

The Lagrangean $L(t,x,\dot{x})$ of a mechanical system is an energy function; in most mechanical systems it is the difference between the kinetic and the potential energy of the system, the kinetic energy being a function convex in $\dot{x}$, while the potential energy depends only on the position x of the system. Thus in the mechanical system consisting of a planet moving around the sun, whose motion was first explained by NEWTON (1687, 2), if (r,θ) denote the polar co-ordinates of the planet with respect to the sun, then the kinetic energy of the system is

$$\frac{m}{2}\left[\dot{r}^2 + r^2\dot{\theta}^2\right]$$

while the potential energy is

$$-g.\frac{mM}{r}$$

where m, M are the masses of the planet and the sun, and g denotes the gravitational constant. The Lagrangean of the planetary system is thus

$$L(r,\theta,\dot{r},\dot{\theta}) = \frac{m}{2}\left[\dot{r}^2 + r^2\dot{\theta}^2\right] + g.\frac{mM}{r}$$

from which LAGRANGE'S equations of motion may be obtained at once.

Thus while in mechanics the motion is characterised by an <u>energy function</u>, and by the way in which a store of energy is generated, in economics the motion is characterised by an objective, or <u>value function</u>, and by the way in which a store of value is generated or used up during the motion. <u>The energy principles that guide the motion of a mechanical system become the value principles that guide the motion of an economic system</u>. Once this transition is understood, the analogy between motion in mechanics and economics will be seen to be most natural.

* * * * *

Chapter IV

On the Duality of Motion: Allocation and Valuation

26. The dynamical theory of valuation originally introduced by IRVING FISHER (1906, 35), depends on two simple principles. The first is the marginal principle of the classical theory of value; according to this principle a commodity is always valued at the margin, by the marginal benefit that an additional unit generates. The value of a commodity to a consumer, for example, is determined by its marginal utility, while the value of commodity to a firm is determined by the value of its marginal product. In the dynamical theory, however, an additional unit of a commodity generates a whole stream of marginal benefit over time.

The second principle, the principle of present value, which has its origin in the method of valuing monetary assets over time, transforms this stream of marginal benefit into a single number. According to this principle, the value of an asset which gives rise to a stream of payments over time is always the present value of the future stream of payments. Thus the value of an asset which gives rise to a stream of $u(t)$ dollars per year for T years, when the annual interest rate is $\delta(t)$, is

$$w(t) = \int_t^T u(\tau) e^{-\int_t^\tau \delta(z)dz} d\tau \qquad (1)$$

Since the value of the asset is obtained from the sum of its future payments, it is clear that its value must fall at each instant at the same rate at which it gives rise to payments, less a certain allowance for appreciation under the force of interest. This result may be obtained immediately by differentiating (1): thus $-\dot{w}(t)$, the rate at which the asset depreciates is given by

$$-\dot{w}(t) = u(t) - \delta w(t) \tag{2}$$

which is merely a differential statement of (1).

As a simple example of the valuation of a monetary asset over time, consider the case of an annuity. In this case $u(t)$ is constant, $u(t) = a > 0$; if the interest rate is also taken as constant, then the value of the annuity at any instant t , is

$$w(t) = a\int_t^T e^{-\delta(\tau - t)}d\tau = \begin{cases} \frac{a}{\delta}(1 - e^{-\delta(T-t)}) & , \ \delta \neq 0 \\ a(T-t) & , \ \delta = 0 \end{cases}$$

and $w(t) = \frac{a}{\delta}$ if the annuity is perpetual. These three cases are shown in Figure 5. Note that if the annuity is finite and $\delta > 0$, then the annuity is always depreciating.

As a second example consider the valuation of a bond. If the bond pays a dollars per year for T years and has a redemption price R , and if the interest rate δ is constant, then the value of the bond at any instant t is

$$w(t) = a\int_t^T e^{-\delta(\tau - t)}d\tau + Re^{-\delta(T-t)} = \frac{a}{\delta} + \left(R - \frac{a}{\delta}\right)e^{-\delta(T-t)}$$

If the redemption price is equal to the par value, so that $a = \alpha R$, where α is the nominal interest rate on the bond, then

$$w(t) = R + (\alpha - \delta)\frac{R}{\delta}(1 - e^{-\delta(T-t)})$$

The value of the bond depends in a simple way on the relation between the nominal and the (market) rate of interest δ. The three possible cases, $\alpha > \delta$, $\alpha = \delta$, $\alpha < \delta$ are shown in Figure 6. When the nominal rate exceeds the market rate the bond is always depreciating: conversely when the market rate exceeds the nominal rate the bond is always appreciating $(t < T)$: when the two rates coincide its value is constant and equal to its par value.

In FISHER'S dynamical theory of valuation, the stream of payments generated by a monetary asset, is replaced by the stream of marginal benefit generated by an additional unit of a commodity. This leads to the following basic principle:

The Valuation Principle.

A commodity is always valued at the margin by the present value of the future stream of marginal benefit that an additional unit of the commodity generates.

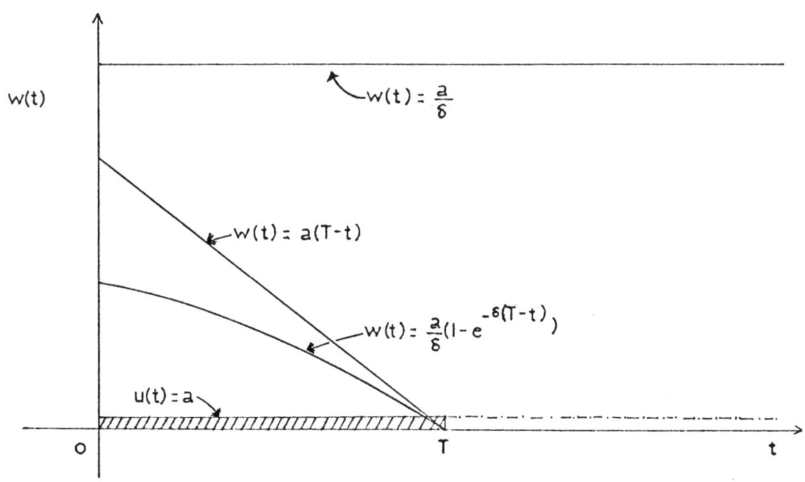

Figure 5. The Value of an annuity.

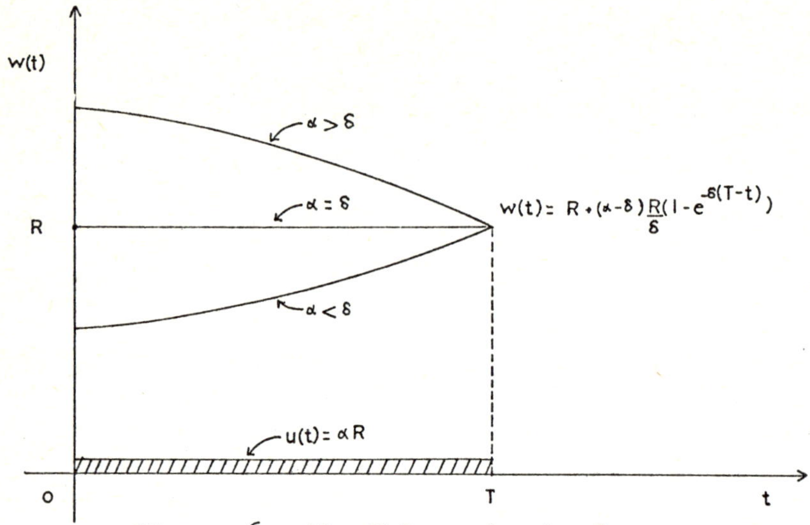

Figure 6. The Value of a bond.

Note the important part played by the future: "the basic problem of time valuation which Nature sets us is always that of translating the future into the present. . ." (1930, 63). The value of a commodity at any instant depends only on the future stream of marginal benefit, the past stream of marginal benefit being of no importance.

It should also be noted, that in this theory the value assigned to a commodity depends on the focal date chosen for the valuation: the focal date is the date in terms of which the valuation is made. Thus if $u(t)$ denotes the stream of marginal benefit generated by an additional unit of a commodity, then the focal date for the value determined by equation (1) is the present, time t, while the focal date for the value

$$v(t) = \int_t^T u(\tau) e^{-\int_0^T \delta(z) dz} d\tau$$

is time zero. Once the valuation in terms of one focal date is known, the valuation in terms of another focal date is easy to obtain: thus if $v(t)$ is known, $w(t)$ is given by

$$w(t) = e^{\int_0^t \delta(z).dz} . v(t)$$

27. In order to understand how this dynamical theory of valuation is embedded in a problem of motion, consider the nature of the decision that needs to be made by an economic system at each instant during the period $[0,T]$. At any given time t, the system has reached a certain position $x(t)$: for fixed t there is nothing the system can do to alter this position. However, during the time-interval

$$[t, t+dt] \tag{3}$$

extending a small distance dt into the future from the fixed time t, the system has the opportunity to change its position by a small amount, say dx, this change in position being the only action that the system can take to influence the benefits that it obtains during the interval of time (3). This change in position or decision, dx, affects the benefits accruing to the system in two principal ways.

The first is the direct effect, namely

$$U(t,x,\frac{dx}{dt}).dt \tag{4}$$

which may therefore be called the direct benefits arising from the decision dx. The second is the indirect effect. In order to determine this second effect, the entire stream of marginal benefit

$$U_x(\tau,x,\dot{x}) \ , \ \tau \in [t, T]$$

generated by the small change in position dx must be known: this requires a knowledge of the trajectory or extremal along which the system is to move; if this trajectory is denoted by E, then the present value of the stream of marginal benefit generated by the decision dx is

$$\left(E\int_t^T U_x(\tau,x,\dot{x}).d\tau\right).dx = \sum_{i=1}^{n}\left(E\int_t^T U_{x_i}(\tau,x,\dot{x}).d\tau\right).dx_i \qquad (5)$$

which may be called the <u>indirect benefits</u> arising from the decision dx . Note that these indirect benefits consist of two parts

$$\left(E\int_t^{t+dt} U_x(\tau,x,\dot{x}).d\tau\right).dx + \left(E\int_{t+dt}^{T} U_x(\tau,x,\dot{x}).d\tau\right).dx$$

namely the current and the future indirect benefits: of these two it is normally the future indirect benefits that are of major importance. The total benefit accruing to the system during the time-interval (3) as a result of the decision dx , is then the sum of these direct and indirect benefits (4) and (5), namely

$$U(t,x,\frac{dx}{dt})dt + \left(E\int_t^T U_x(\tau,x,\dot{x}).d\tau\right).dx \qquad (6)$$

In view of IRVING FISHER'S Valuation Principle the integral in (6) induces a vector of valuations of each of the co-ordinates $x_1,\dots,x_n$ whose components

$$q_i = E\int_t^T U_{x_i}(\tau,x,\dot{x}).d\tau \ , \ i=1,\dots,n \qquad (7)$$

will be called the <u>imputed values</u> of the co-ordinates $x_1,\dots,x_n$. The total benefits (6) accruing to the system may then be written in terms of these imputed values,

$$U(t,x,\frac{dx}{dt}).dt + q\,dx \qquad (8)$$

and will be called the total imputed value generated by the decision dx during the time-interval $[t, t+dt]$.

Dividing (8) by dt gives the flow of total imputed value generated by the decision $\dot{x} = \frac{dx}{dt}$ at the instant t , namely

$$U(t,x,\dot{x}) + q\dot{x} \qquad (9)$$

The above analysis thus leads to the following basic Principle of Motion which replaces the Principle of Stationary Present Value in the dual theory of motion:

The Maximum Principle.

If the trajectory of an economic system is determined by the condition that its Present Value attain a maximum, then the decision $\dot{x}$ must be chosen at each instant t during the time-interval $[0,T]$ so as to maximise the (flow of) total imputed value (9) generated by the system at that instant.

The maximum value of (9) depends on the imputed value q and leads to the function

$$H(t,x,q) = \max_{\dot{x}} \{U(t,x,\dot{x}) + q\dot{x}\} \qquad (10)$$

which may be called the dual objective function since the objective of the combined problem of valuation and allocation dual to the original pure allocation problem is defined by this function. If the maximum in (10) is to be attained then $\dot{x}$ must be chosen so that

$$U_{\dot{x}_i} + q_i = 0 \qquad i = 1,\dots,n . \qquad (11)$$

(11) has a simple economic interpretation. For it implies that <u>the indirect marginal benefit q of a increase in $\dot{x}$ must equal the direct marginal opportunity-cost $-U_{\dot{x}}$.</u> (11) is thus the <u>marginal condition</u> that must hold if the Maximum Principle is to be satisfied.

If $U(t,x,\dot{x})$ is not only concave, but strictly concave in $\dot{x}$, so that the quadratic form

$$\sum_{i,j=1}^{n} U_{\dot{x}_i \dot{x}_j} \alpha_i \alpha_j$$

is negative definite, then the maximal decision $\dot{x}$ may be otained from (11) by inversion:

$$\dot{x}_i = \varphi_i(t,x,q) \quad , \quad i = 1,\ldots,n$$

the dual objective function thus becomes

$$H(t,x,q) = U(t,x,\varphi(t,x,q)) + q\varphi(t,x,q) \tag{10}$$

Differentiating H with respect to q_i gives

$$Hq_i = \sum_{j=1}^{n} (U_{\dot{x}_j} + q_j)\varphi_{j_{q_i}} + \varphi_i = \dot{x}_i \quad , \quad i = 1,\ldots,n \tag{12}$$

similarily, differentiating with respect to x_i gives

$$Hx_i = U_{x_i} + \sum_{j=1}^{n} (U_{\dot{x}_j} + q_j)\varphi_{j_{x_i}} = U_{x_i} \quad , \quad i = 1,\ldots,n \tag{13}$$

by virtue of (11). Finally differentiating (7) and using (13) implies that

$$\dot{q}_i = -U_{x_i} = -H_{x_i} \quad , \quad i = 1,\ldots,n \tag{14}$$

which is merely the differential statement of the Valuation Principle contained in (7) and is equivalent to the earlier equation

(2). (14) therefore has a simple economic interpretation. For it implies that each co-ordinate $x_1, \ldots, x_n$ of the economic system depreciates at the same rate at which it gives rise to marginal benefits.

(12) and (14) are the allocation and valuation equations of the dual theory of motion and together may be called the dual equations of motion, since they determine not only the trajectory $x(t)$ of the economic system, but also the trajectory of the dual imputed value $q(t)$.

Note that while the definition of the imputed values contained in (7) required a knowledge of the actual trajectory E of the system, on differentiating (7) to obtain (14), a knowledge of this trajectory is no longer required. The construction of the dual theory of motion does thus not assume a knowledge of any information that is unavailable to the economic system.

A simple economic interpretation may be given of the manner in which the dual equations of motion guide the economic system along its trajectory. Consider the begining of the time-interval $[0,T]$. Given $x(0)$ and $q(0)$, the Maximum Principle, and hence the equations (11) or (12), lead to the best initial decision $\dot{x}(0)$: this determines the position at a subsequent instant. In the meantime, the Valuation Principle implies that the imputed value q depreciates by an amount equal to the value of the marginal benefit, H_{x_i}, that it has yielded: this determines a new value for q at the subsequent instant. The situation at time zero now repeats itself, with the new values of x and q replacing $x(0)$ and $q(0)$, $\dot{x}$ being chosen to satisfy the Maximum Principle, and $\dot{q}$ obeying FISHER'S Valuation Principle. At each instant the allocation equations (12) embody the Maximum Principle, while the valuation equations (14) embody the Valuation Principle. These two principles together guide the economic system along its trajectory of motion.

It is not difficult to see that the conditions implied by the Maximum Principle and the Valuation Principle, contained in (11) and (14) replace the conditions required by the pure allocation equation of motion (3) of Chapter III.

28. This dual theory of motion may be given a simple and elegant geometrical interpretation. For at any fixed instant t, the economic system has reached a certain position $x(t)$; the objective function $U(t,x,\dot{x})$, with t and x held at these fixed values, defines a surface

$$U = U(\dot{x})$$

called the indicatrix, in the $(n+1)$ dimensional decision space

$$\mathfrak{U} = \{(U,\dot{x})\} .$$

As indicated in Chapter III, the indicatrix in turn generates a convex region $\mathcal{J} \subset \mathfrak{U}$

$$\mathcal{J} = \{(U,\dot{x}) \mid U \leq U(\dot{x}) , \dot{x} \in D\}$$

where D is the convex domain in $\mathbb{R}^n$ on which $U(\dot{x})$ is defined and where $U(\dot{x})$ is now taken as strictly concave in $\dot{x}$. At the instant t, the vector of imputed values q, which is assumed to be known, leads to a family of parallel hyperplanes in $\mathfrak{U}$,

$$U + q\dot{x} = h \qquad (15)$$

depending on the parameter $h \geq 0$.

The geometrical statement of the Maximum Principle is then as follows: the point $(U,\dot{x}) \in \mathfrak{U}$ must be chosen from among the points lying in the convex feasible region $\mathcal{J} \subset \mathfrak{U}$, so as to lie on the hyperplane (15) that is furthest removed from the origin.

Such a point always exists provided that q lies in a certain convex region $C \subset \mathbb{E}^n$, which depends on the form of the convex region $\mathcal{J}$, for $\mathcal{J}$ is not necessarily a bounded convex region of $\mathfrak{U}$.

The transformation of the problem into dual form thus reduces the problem of motion to the static problem of choosing the optimal position $(U,\dot{x}) \in \mathfrak{U}$ at each instant. This optimal point $(U,\dot{x}) \in \mathfrak{U}$ is determined by the condition that the hyperplane (15) must be tangent to the indicatrix. This is the geometrical interpretation of the marginal condition (11).. For the tangency condition implies that the gradients of the two surfaces must be linearly dependent: since the gradients are $(1,q)$ and $(-1,U_{\dot{x}})$, (11) must be satisfied. This point may also be thought of as being determined by NEWTON'S principle that the sum of the forces $(1,q)$ and $(-1,U_{\dot{x}})$ acting on the system must be zero.

The maximum value of the parameter h attained by the hyperplane (15) at the optimal point, depends on the imputed value q and may be denoted by $H(q)$ as in (10), thus

$$H(q) = \max_{(U,\dot{x}) \in \mathcal{J}} \{U + q\dot{x}\}$$

Since for every point $(U,\dot{x}) \in \mathcal{J}$, the inequality

$$U + q\dot{x} \leq H(q)$$

is satisfied, while the optimal point $(U,\dot{x}) \in \mathcal{J}$ satisfies

$$U + q\dot{x} = H(q)$$

$H(q)$ is a <u>supporting hyperplane</u> for the convex region $\mathcal{J}$. Since $H(q)$ generates a whole family of supporting hyperplanes for $\mathcal{J}$ as q ranges through the convex domain C, $H(q)$ is called <u>the function of support of the convex region $\mathcal{J}$</u>. This function, which was first introduced by MINKOWSKI (1911, 41), completely characterises the convex region $\mathcal{J}$, for $\mathcal{J}$ is the collection of all points $(U,\dot{x}) \in \mathfrak{U}$ satisfying

$$U + q\dot{x} \leq H(q) \quad \text{for all} \quad q \in C . \tag{16}$$

(11) and (12) imply that

$$\dot{x}_i = H_{q_i}(t,x,q) = H_{q_i}(t,x,-U_{\dot{x}}) \quad i = 1,\ldots,n$$

differentiating with respect to $\dot{x}_k$ gives

$$\delta_{ik} = -\sum_{j=1}^{n} H_{q_i q_j} U_{\dot{x}_j \dot{x}_k} \qquad i,k = 1,\ldots,n$$

where $\delta_{ik} = 0$ for $i \neq k$, and $\delta_{ik} = 1$ for $i = k$. Introducing the variables

$$\beta_j = \sum_{k=1}^{n} U_{\dot{x}_j \dot{x}_k} \alpha_k \ , \ j = 1,\ldots,n$$

multiplying by $H_{q_i q_j}$ and summing, gives

$$\sum_{j=1}^{n} \beta_j H_{q_i q_j} = \sum_{j,k=1}^{n} H_{q_i q_j} U_{\dot{x}_j \dot{x}_k} \alpha_k = -\delta_{ik}\alpha_k = -\alpha_i \quad , \quad i = 1,\ldots,n$$

which implies that

$$\sum_{j=1}^{n} \beta_j \alpha_j = \sum_{k,j=1}^{n} U_{\dot{x}_j \dot{x}_k} \alpha_k \alpha_j = -\sum_{i,j=1}^{n} H_{q_j q_i} \beta_i \beta_j$$

<u>Thus if $U(\dot{x})$ is a strictly concave function, then $H(q)$ is a strictly convex function</u>.

Now consider the hyperplanes (15) defined on the decision space $\mathfrak{U}$. The coefficients of these hyperplanes are the covectors of the space

$$\mathfrak{U}^* = \{(H,q)\}$$

dual to the decision space $\mathfrak{U}$, which may be called the <u>imputed value space</u>. Just as the objective function $U(\dot{x})$ defines a concave surface $U = U(\dot{x})$ in $\mathfrak{U}$, called the indicatrix, which in turn leads to the convex region $\mathcal{J} \subset \mathfrak{U}$, so the dual objective function $H(q)$ defines a convex surface

$$H = H(q)$$

in $\mathfrak{U}^*$, called the <u>figuratrix</u>, which in turn generates the convex region $\mathcal{J}^* \subset \mathfrak{U}^*$

$$\mathcal{J}^* = \{(H,q) \mid H \geq H(q) \ , \ q \in C\}$$

which is dual to the convex region $\mathcal{J} \subset \mathfrak{U}$.

While in the formulation of the Maximum Principle in $\mathfrak{U}$, it is assumed that at each instant t the imputed value q is known, and the maximal decision $\dot{x}$ is determined given the imputed value q , in the dual space $\mathfrak{U}^*$ the situation is reversed: the decision $\dot{x}$ is assumed to be known, and a dual <u>Minimum Principle</u> is used to determine the minimal imputed value q given the decision $\dot{x}$. According to this Minimum Principle, given the decision $\dot{x}$ at each instant, <u>the imputed value q must be chosen so as to minimise the current direct benefits $H(q) - \dot{x}q$</u> .

Just as the Maximum Principle has a simple geometrical interpretation in $\mathfrak{U}$, so the Minimum Principle has a simple geometrical interpretation in $\mathfrak{U}^*$. For at each instant, the decision vector

$\dot{x}$, which is assumed to be known, leads to a family of parallel hyperplanes in $\mathfrak{U}^*$

$$H - \dot{x}q = u \tag{17}$$

depending on the parameter $u \geq 0$. The Minimum Principle then implies that among the points lying in the convex feasible region $\mathcal{J}^* \subset \mathfrak{U}^*$, the point (H,q) must be chosen, which lies on the hyperplane (17) that is nearest to the origin.

Such a point always exists provided that $\dot{x}$ lies in the convex region D . The minimal point $(H,q) \in \mathcal{J}^*$ is then determined by the condition that the hyperplane (17) must be tangent to the figuratrix. This is the geometrical interpretation of the marginal condition (12). For the tangency condition implies that the gradients of the two surfaces must be linearly dependent: since the gradients are $(1,-\dot{x})$ and $(-1,H_q)$, (12) must be satisfied.

The minimum value of the parameter u attained by the hyperplane (17) at the minimal point, depends on the decision $\dot{x}$. (16) implies that this minimum value is $U(\dot{x})$, thus

$$U(\dot{x}) = \min_{(H,q) \in \mathcal{J}^*} \{H - \dot{x}q\}$$

Since for every point $(H,q) \in \mathcal{J}^*$ the inequality

$$H - \dot{x}q \geq U(\dot{x})$$

is satisfied, while the minimal point $(H,q) \in \mathcal{J}^*$ satisfies

$$H - \dot{x}q = U(\dot{x})$$

$U(\dot{x})$ is a supporting hyperplane for the convex region $\mathcal{J}^*$. Since $U(\dot{x})$ generates a whole family of supporting hyperplanes for $\mathcal{J}^*$, as $\dot{x}$ ranges through the convex domain D , $U(\dot{x})$ is the function of the convex region $\mathcal{J}^*$. The convex region $\mathcal{J}^*$ may then be characterised in terms of its function of support $U(\dot{x})$, as the collection of all points $(H,q) \in \mathcal{U}^*$ satisfying

$$H - \dot{x}q \geq U(\dot{x}) \quad \text{for all} \quad \dot{x} \in D .$$

There is thus a completely symmetrical relation between the objective function, the dual objective function and the two convex regions which they generate, $H(q)$ acting as the function of support of the convex region $\mathcal{J}$ generated by $U(\dot{x})$, and $U(\dot{x})$ acting as the function of support of the convex region $\mathcal{J}^*$ generated by $H(q)$.

29. This geometric interpretation of the dual theory may be illustrated briefly by the second example of Chapter III. Thus if $u(c) = \sqrt{c}$, then the objective function (the indicatrix) becomes

$$U(\dot{k}) = e^{-\delta t}\sqrt{f(k) - \lambda k - \dot{k}}$$

Since at any instant t , investment cannot exceed net output, while disinvestment cannot exceed the total capital stock in existence,

$$-k(t) \leq \dot{k} \leq f(k) - \lambda k$$

the interval on which $U(\dot{k})$ is defined is therefore $D = [-k(t) , f(k) - \lambda k]$. The objective function and the convex region $\mathcal{J}$ of the decision space are shown in Figure 7.

Using the marginal condition (11) to determine the optimal investment $\dot{k}$ as a function of the imputed value of capital q , leads to the dual objective function (10)′ (the figuratrix),

$$H(q) = \frac{e^{-2\delta t}}{4q} + (f(k) - \lambda k)q$$

If $-q_m$ denotes the slope of the indicatrix when $\dot{k} = -k(t)$, since its slope is infinite when $\dot{k} = f(k) - \lambda k$, the interval on which

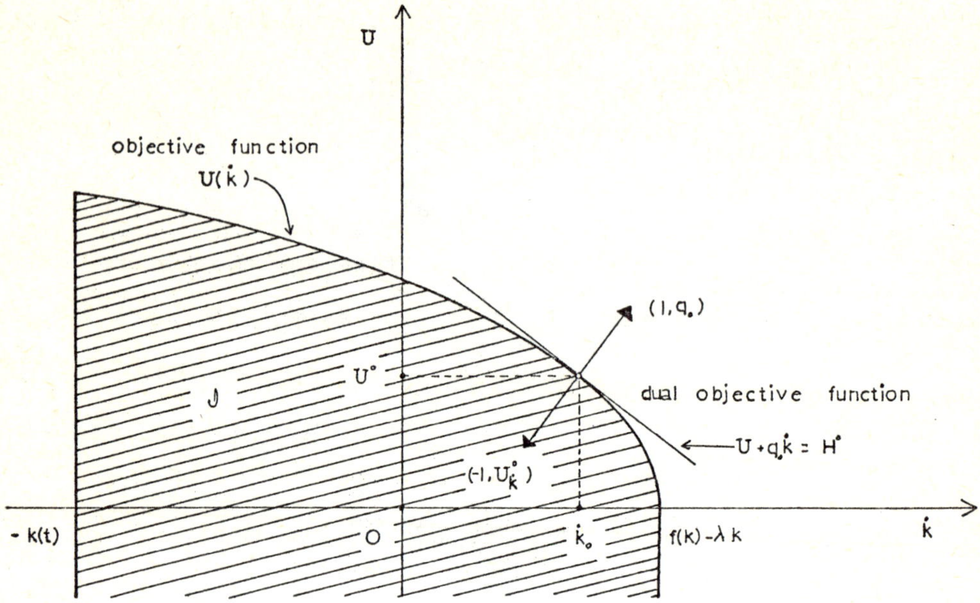

Figure 7. The dual objective function is the supporting plane of of the convex region $\mathcal{J}$ defined by the objective function.

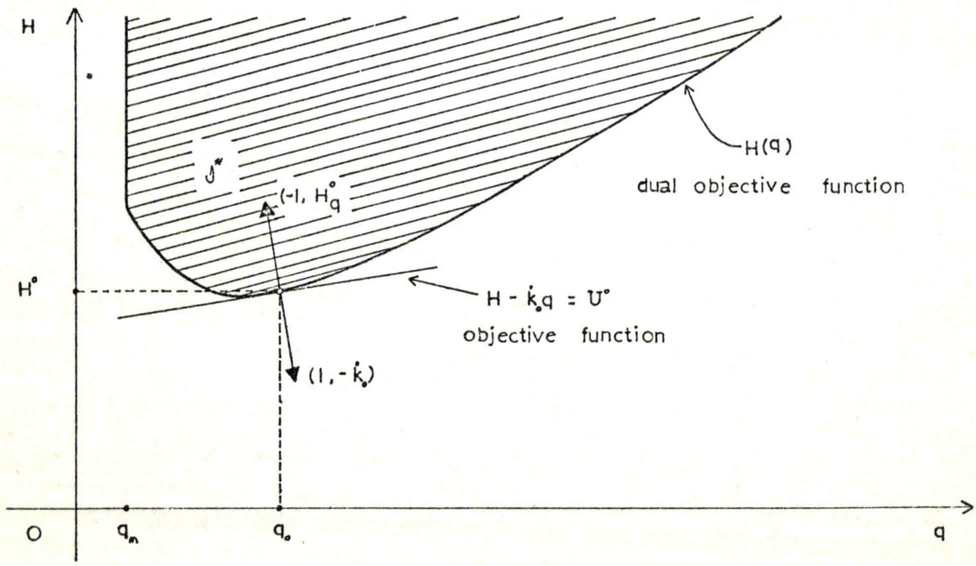

Figure 8. The objective function is the supporting plane of the convex region $\mathcal{J}^*$ defined by the dual objective function.

$H(q)$ is defined as $C = [q_m, \infty]$. The dual objective function and the dual convex region $\mathcal{J}^*$ of the imputed value space are shown in Figure 8.

Consider the relation between investment and the imputed value of a unit of capital implied by Figures 7 and 8. If the negative slope of the supporting plane for $\mathcal{J}$ increases, the optimal amount of investment $\dot{k}$ increases: thus if the imputed value of capital increases, the optimal amount of investment increases. Since the slope of the supporting plane is infinite when $\dot{k} = f(k) - \lambda k$, provided the imputed value is always finite, investment never absorbs all of net output. Since the slope of the supporting plane for $\mathcal{J}^*$ denotes the investment under taken and since $H(q)$ attains a minimum for

$$\bar{q} = \sqrt{\frac{e^{-2\delta t}}{4(f(k) - \lambda k)}}$$

investment is negative whenever $q < \bar{q}$, zero for $q = \bar{q}$ and positive for $q > \bar{q}$.

$(U^o, \dot{k}_o)$ is the optimal point in the decision space $\mathcal{U}$ associated with the imputed value q_o : conversely (H^o, q_o) is the optimal point in the imputed value space $\mathcal{U}^*$ associated with the investment $\dot{k}_o$. Note the relation between these two points $(U^o, \dot{k}_o) \in \mathcal{U}$ and $(H^o, q_o) \in \mathcal{U}^*$. q_o and H^o are the coefficients of the supporting plane for $\mathcal{J}$ at $(U^o, \dot{k}_o)$, while $\dot{k}_o$ and U^o are the coefficients of the supporting plane for $\mathcal{J}^*$ at (H^o, q_o) . Thus the boundary points of $\mathcal{J}^*$ are generated by the coefficients of the supporting planes of $\mathcal{J}$, while the boundary points of $\mathcal{J}$ are generated by the coefficients of the supporting planes of $\mathcal{J}^*$. This is the geometric interpretation of the one-to-one map between

$\mathfrak{U}$ and $\mathfrak{U}^*$ defined implicitly by the <u>LEGENDRE transformation</u>

$$q + U_{\dot{x}} = 0 \quad , \quad H = U + q\dot{x}$$

and its inverse

$$\dot{x} - H_q = 0 \quad , \quad U = H - \dot{x}q \; .$$

30. The dual theory of motion may be unified in a very remarkable way, by introducing a function which represents the analogue in economic theory of the Characteristic Function first introduced by HAMILTON into optics (1824, 8), and the Principal Function which he subsequently introduced into analytical mechanics (1834, 6; 1835, 7).

In order to derive this function in a natural manner consider LAGRANGE'S expression (1760, 64) for the general variation δI of the functional

$$I = \int_t^{t'} U(\tau, x, \dot{x}).d\tau$$

namely

$$\delta I = \int_t^{t'} \sum_{i=1}^{n} \left(U_{x_i} - \frac{d}{d\tau}(U_{\dot{x}_i}) \right) h_i(\tau).d\tau$$

$$+ \left[\sum_{i=1}^{n} U_{\dot{x}_i} \delta_{x_i} + (U - \sum_{i=1}^{n} \dot{x}_i U_{\dot{x}_i}) \delta\tau \right]_t^{t'} \qquad (18)$$

If the trajectory joining the two end-points $(t, x(t))$ and $(t', x(t'))$ is the actual trajectory of the economic system then the equations of motion (3) of Chapter III are satisfied and the integral in (18) vanishes. The variation is then a function of the end-points alone; this leads at once to HAMILTON'S <u>Principal Function</u>

$$R(t,x(t),t',x(t')) = {}_E\int_t^{t'} U(\tau,x,\dot{x}).d\tau$$

where the integral is evaluated along the actual trajectory or extremal E joining the end-points. Substituting (10)'and (11) into (18) implies

$$dR = [-qdx + Hd\tau]_t^{t'}$$

thus

$$\begin{aligned} R_{t'} &= H(t') \ , \ R_{x(t')} = -q(t') \\ R_t &= -H(t) \ , \ R_{x(t)} = q(t) \end{aligned} \qquad (19)$$

which imply

$$\begin{aligned} R_{t'} - H(t',x(t'), -R_{x(t')}) &= 0 \\ R_t + H(t,x(t),R_{x(t)}) &= 0 \end{aligned} \qquad (20)$$

which are HAMILTON'S two partial differential equations for the final and initial co-ordinates $(t',x(t'))$ and $(t,x(t))$ of the Principal Function. These two equations are not independent however, as JACOBI demonstrated (1866, 65), for a solution of the equation in the final co-ordinates may always be obtained from a solution of the equation in the initial co-ordinates. This is equivalent to holding the final co-ordinates of the Principal Function fixed, and allowing only the initial co-ordinates to vary, an observation which leads at once to a valuation of the economic system as a whole similar to the valuations (7) introduced earlier for each of the co-ordinates; for by holding the final co-ordinates of the Principal Function fixed and letting $(t',x(t') = (T,x(T))$ we obtain the function

$$V(t,x) = E\int_t^T U(\tau,x,\dot{x}).d\tau$$

which may be called the Present Value Function of the economic system, for $V(t,x)$ represents at each instant the present value of the future stream of benefits that will accrue to the economic system.

The vector of imputed values q, as well as the total imputed value H, may be obtained immediately from the Present Value Function, for (19) implies

$$q = V_x$$

$$H = -V_t$$

The first equation also follows at once from the defintion (7) of the imputed value q. While q is the vector of valuations induced by the co-ordinates, H, the total imputed value, is the valuation induced by time; for $-V_t$ denotes the marginal benefit obtained from a small increase in time, or the marginal benefit sacrificed by a small decrease in time. This benefit is exactly the total imputed value generated during a marginal unit of time.

(20) implies that the Present Value Function satisfies the HAMILTON-JACOBI equation

$$V_t + H(t,x,V_x) = 0 \qquad (20)$$

which is the fundamental valuation equation characterising the motion of the economic system. This valuation equation embodies a more general statement of the process underlying the motion than the dual equations (12) and (14), for the characteristic equations

$$\frac{dt}{1} = \frac{dx_1}{H_{q_1}} = \dots = \frac{dx_n}{H_{q_n}} = \frac{dq_1}{-H_{x_1}} = \dots = \frac{dq_n}{-H_{x_n}}$$

of (20) are equivalent to these earlier equations. Indeed when a complete integral $V(t,x,a)$ of (20) can be found, the solution of the dual equations may be obtained by differentiation and elimination from the equations

$$V_a = b$$

$$V_x = q$$

where the vectors of constants a and b are determined by the initial conditions, as was shown by JACOBI (1866, 65). The principal importance of the valuation equation (20) however, derives from the relation that it generates between the surfaces of constant Present Value and the trajectories of motion, a relation which is examined in Chapter V.

31. The dual theory of motion developed so two may be illustrated briefly by the two examples introduced earlier.

Consider first the example of the corporation. At any instant t during the period $[0,T]$, the corporation has at its disposal a certain positive amount of permanent capital $k(t)$. However, since this permanent capital is limited, only a limited amount of profit $f(k(t))$ can be obtained, of which an amount $\alpha\beta k(t)$ must be used to pay interest charges on the debt. There remains an amount $f(k(t)) - \alpha\beta k(t)$, which may be called the net (short-term) profit of the corporation. The managers are now faced with a problem of choice: for the limited net profit has two alternative uses: the net profit may either be distributed as dividends to the shareholders or it may be retained, and used to increase the permanent capital of the corporation.

In order to decide how the limited net profit may best be allocated between dividends and investment, the managers must know the benefits that accrue from each of these two uses. In particular they must know the direct marginal opportunity cost and the indirect marginal benefit arising from an additional unit of investment, for an increase in investment implies an equal amount of dividends foregone. Since an additional unit of investment automatically provides β of a unit of funds through borrowing, only $(1-\beta)$ of a unit of dividends needs to be sacrificed to provide an additional unit of investment. The current direct marginal opportunity-cost of an additional unit of investment is thus $(1-\beta)\,u'(d)$. On the other hand, from the valuation theory it is known that the indirect marginal benefit of an additional unit of investment is $q(t)$, the focal date being at time zero. In order to compare the indirect marginal benefit and the direct marginal cost the focal date for $q(t)$ must be transformed to the present; this may be done by introducing the imputed value $p(t)=q(t)e^{\delta t}$.

Now the marginal condition (11) must be satisfied if the managers are to obtain maximum benefit from the currently available net profit. The optimal amount of investment is thus determined by the equation

$$q=-U_{\dot{k}}=(1-\beta)u'\,(f(k)-\alpha\beta k-\dot{k}(1-\beta))e^{-\delta t} \qquad (21)$$

or

$$p=(1-\beta)u'\,(d)$$

<u>The indirect marginal benefit from an additional unit of investment must equal its direct marginal opportunity-cost.</u>

Solving (21) for the optimal investment gives

$$\dot{k}=\frac{1}{1-\beta}\left[f(k)-\alpha\beta k-u'^{-1}\left(\frac{qe^{\delta t}}{1-\beta}\right)\right]$$

thus the dual objective function becomes

$$H(t,x,q) = u(u'^{-1}(\frac{qe^{\delta t}}{1-\beta}))e^{-\delta t} + \frac{q}{1-\beta}\left[f(k) - \alpha\beta k - u'^{-1}(\frac{qe^{\delta t}}{1-\beta})\right]$$

which in turn leads to the valuation equation

$$\dot{q} = -H_k = -\frac{q}{1-\beta}\left[f'(k) - \alpha\beta\right]$$

Transforming the focal date for the valuation to the present leads to the dual equations of motion

$$\dot{k} = \frac{1}{1-\beta}\left[f(k) - \alpha\beta k - u'^{-1}(\frac{p}{1-\beta})\right]$$

$$\dot{p} = -\frac{p}{1-\beta}\left[f'(k) - \alpha\beta - \delta(1-\beta)\right] \qquad (22)$$

where the latter equation determines the manner in which the valuation of the permanent capital changes over time.

If the focal dates for the dual objective function and the Present Value Function are transformed to the present by introducing the functions

$$G(k,p) = e^{\delta t}H(t,x,q)$$

$$W(k,t) = e^{\delta t}V(k,t)$$

then the basic valuation equation (20) becomes

$$W_t + G(k,W_k) - \delta W = 0$$

which implies that $W(k,t)$ satisfies the equation

$$W_t + u(u'^{-1}(\frac{W_k}{1-\beta})) + \frac{W_k}{1-\beta}\left[f(k) - \alpha\beta k - u'^{-1}(\frac{W_k}{1-\beta})\right] - \delta W = 0 \qquad (23)$$

When the exact forms of the functions $f(k)$ and $u(d)$ remain unspecified, however, as in the present analysis, it is easier to analyse the motion in terms of the dual equations (22) than the valuation equation (23).

The dual equations (22) are the basic equations which determine the investment undertaken at each instant. The geometric form of the solutions to these equations may be determined by using the method of POINCARÉ introduced in Chapter III. Thus $\dot{p} = 0$, $p \neq 0$, implies $f'(k) - \alpha\beta - \delta(1-\beta) = 0$ or $k = \hat{k}$. $\dot{k} = 0$ when $f(k) - \alpha\beta k - u'^{-1}(\frac{p}{1-\beta}) = 0$ or $p = (1-\beta)u'(f(k) - \alpha\beta k)$. Note that the slope of this curve is

$$\frac{dp}{dk} = (1-\beta)u''(f(k) - \alpha\beta k)\left[f'(k) - \alpha\beta\right] = \begin{cases} 0 \ , & k = \bar{k} \\ >0 \ , & k > \bar{k} \\ <0 \ , & k < \bar{k} \end{cases}$$

where $\bar{k}$ is defined by $f'(\bar{k}) - \alpha\beta = 0$. $\dot{k} = 0$, $\dot{p} = 0$ simultaneously when $k = \hat{k}$, $p = \hat{p} = (1-\beta)u'(f(\hat{k}) - \alpha\beta\hat{k})$. Since the roots of the linear approximation to (22) around $(\hat{k},\hat{p})$ are

$$r_1, r_2 = \frac{1}{2}\left[\delta \pm \sqrt{\delta^2 + \frac{4\hat{p}f''(\hat{k})}{(1-\beta)^2 u''(\hat{d})}}\right]$$

$r_1 > 0$, $r_2 < 0$ as before and the solution curves form a saddlepoint around $(\hat{k},\hat{p})$ as indicated in Figure 9.

The analysis of the motion now proceeds in a manner similar to that in Chapter III, reducing to the consideration of the motion of the two main kinds of corporation. Figure 9 reveals the way in which the imputed value $p(t)$ changes for the two main kinds of corporation. In the first kind, given $k(0)$, $p_1(0)$ is immediately determined: in the second kind, the growth-orientated corporation, $p_2(0)$ is obtained from $k(0)$ and $k_2(T)$. Given $(p_1(0),k(0))$ and $(p_2(0),k(0))$, the dual equations then determine

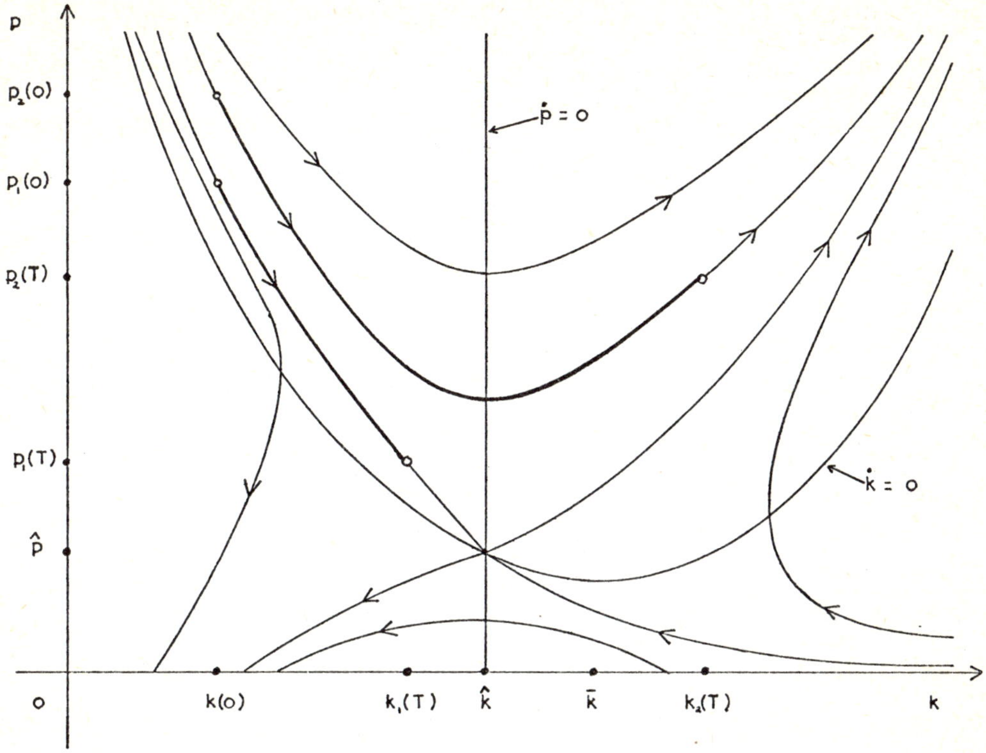

Figure 9. Dual Motion of the corporation

the way in which the corporation grows over time at each subsequent instant, as shown in Figure 9.

32. As a second example consider the dual formulation of the motion of the aggregate economy.

The Planning Authority faces a problem of choice at each instant, similar to that faced by the managers of a corporation. For the economy is endowed at each instant with a limited amount of capital-per-worker $k(t)$, from which only a limited output-per-worker $f(k(t))$ can be generated. (In future, output, capital, investment and consumption will all be understood to be measured in amounts per-worker.) After $\lambda k(t)$ has been deducted for the replacement of worn-out capital equipment, the Planning Authority is left with the <u>net output</u> $f(k(t)) - \lambda k(t)$, which may be allocated either towards consumption by the members of the community or towards investment.

The optimal amount of investment is determined by the marginal condition (11); thus

$$q = -U_{\dot{k}} = u'(f(k) - \lambda k - \dot{k})e^{-\delta t}$$

where q is the imputed value of capital. Transforming the focal date to the present, by introducing the imputed value $p = qe^{-\delta t}$, implies

$$p = u'(c)$$

which is the basic <u>welfare equation</u> [Lerner (1944, 66)] that must be satisfied, if the allocation of net output between consumption and investment is to be optimal: <u>the marginal social benefit of an additional unit of investment must equal its marginal social opportunity cost, the marginal benefit of the unit of consumption foregone</u>.

The dual equations

$$\begin{aligned} \dot{k} &= f(k) - \lambda k - u'^{-1}(p) \\ \dot{p} &= -p(f'(k) - (\delta + \lambda)) \end{aligned} \tag{24}$$

and the valuation equation

$$W_t + u(u'^{-1}(W_k)) + W_k(f(k) - \lambda k - u'^{-1}(W_k)) - \delta W = 0$$

may then be derived exactly as in the previous example. An analysis of the dual equations (24) reveals a similar catenary motion around the critical point (k^*, p^*) , where $p^* = u'(f(k^*) - \lambda k^*)$.

33. It is perhaps appropriate to recall the similarity between the present dual theory of motion, and the canonical theory of motion

in analytical mechanics. For the transformation from the pure allocation equations of motion to the dual equations of motion in economic theory, corresponds to the transformation from LAGRANGE'S equations to the canonical equations of motion in analytical mechanics. In analytical mechanics however, the vector of conjugate momenta $p = (p_1, \ldots, p_n)$ and the Hamiltonian H, which correspond to the vector of imputed values and the dual objective function, are defined by a LEGENDRE transformation different from that in economic theory, for Nature chooses the velocity vector $\dot{x}$ at each instant so that the total energy of the system

$$p\dot{x} - L$$

attains a maximum. The LEGENDRE transformation thus becomes

$$p = L_{\dot{x}} \; , \; H = p\dot{x} - L \tag{25}$$

Just as an economic system may be characterised by its Present Value Function, so a mechanical system may be characterised by its Action Function $S(t',x)$. There is however, a fundamental difference between these two functions, and hence in the nature of the two process which they represent. For the Present Value Function $V(t,x)$, is a function which depends on a store of value to be generated in the future, while the Action Function $S(t',x)$, is a function which depends on the store of energy that has been generated in the past. Thus while $V(t,x)$ is obtained from the Principal Function by varying the intial co-ordinates, and holding the final co-ordinates fixed, the Action Function $S(t',x)$ is obtained from the Principal Function by holding the initial co-ordinates, which may be written as $(0,x(0))$, fixed, and varying the final co-ordinates $(t',x(t'))$. Thus

$$S(t',x) = E\int_{0}^{t'} L(\tau,x,\dot{x}).d\tau$$

(18) and the LEGENDRE transformation (25) thus imply that the Action Function satisfies the HAMILTON-JACOBI equation

$$S_{t'} + H(t',x,S_x) = 0$$

which is exactly the same as the valuation equation (20) satisfied by the Present Value Function.

Since

$$p = S_x = E\int_{0}^{t'} L_x(\tau,x,\dot{x}).d\tau$$

$$\dot{p} = L_x = -H_x$$

and since (25) implies $\dot{x} = H_p$, the canonical equations of motion of the mechanical system become

$$\dot{x} = H_p$$
$$\dot{p} = -H_x$$

These equations, which were first formally introduced by HAMILTON (1835, 7), are exactly the same as the dual equations of motion (12) and (14) of an economic system.

As in Chapter III, the example of the planetary system may be used to illustrate the dual formulation of the theory in analytical mechanics; (25) implies that the Hamiltonian for the system is

$$H(t,x,p) = \frac{1}{2m}\left(\left(\frac{p_\theta}{r}\right)^2 + p_r^2\right) - \frac{gmM}{r}$$

from which the canonical equations of motion, and the HAMILTON-JACOBI equation are easily derived.

Thus while the dual formulation of the theory of motion is essentially the same in analytical mechanics and in economics, there is one major distinction between the two theories: for motion in economic theory depends essentially on the future, while motion in mechanics depends essentially on the past. This basic distinction is embodied in the two fundamental functions that characterise the motion of the systems in the two theories, namely the Present Value Function and the Action Function. From this basic difference arises the distinction between the LEGENDRE transformations in the two theories, for an examination of the fundamental variational expression (18) reveals that these are the two transformations from $\mathfrak{U}$ into $\mathfrak{U}^*$ that yield the desired positive conjugate momenta and Hamiltonian in analytical mechanics and the desired positive imputed values and dual objective function in economics.

* * * * *

Chapter V

On the Possibility of Wave Motion

34. The Hamilton-Jacobi equation reveals a very elegant structure underlying the economic theory of motion. For the Hamilton-Jacobi equation embodies Huygens Principle of wave propagation. The surfaces of constant Present Value are thus propagated in the same way as the wave-fronts in the theory of light. The analysis reveals a remarkable parallel between the economic theory of motion and the theory of the propagation of light.

For suppose that in the economic theory of motion the integral

$$E\int_{(x_1,t_1)}^{(x_2,t_2)} U(\tau,x,\dot{x}).d\tau$$

denotes the <u>distance</u> between the two points (x_1,t_1) and (x_2,t_2) in the $n+1$ dimensional space $\mathbb{E}^{n+1}$. Consider a family of surfaces $S(x,t)=\alpha$ defined on $\mathbb{E}^{n+1}$. Such a family of surfaces will be called a <u>parallel family</u> of surfaces if for any two surfaces $S(x,t)=\alpha_1$ and $S(x,t)=\alpha_2$ of the family, where $\alpha_1>\alpha_2$,

$$E\int_{(x_1,t_1)}^{(x_2,t_2)} U(\tau,x,\dot{x}).d\tau=\alpha_1-\alpha_2$$

where (x_2,t_2) is any point on the surface $S(x,t)=\alpha_2$, and (x_1,t_1) is the point of intersection of the extremal passing through (x_2,t_2) with the surface $S(x,t)=\alpha_1$. The parallel surfaces in E^{n+1} can then be given the following simple characterisation: <u>a family of surfaces $S(x,t)=\alpha$ is parallel if and only if each member satisfies the Hamilton-Jacobi equation $S_t+H(t,x,S_x)=0$</u> [Rund (1966, 67)]. Thus if the objective function is used as a measure of distance then the surfaces of constant Present Value, which are solutions of the Hamilton-Jacobi equation, mark off equal distances along the trajectories of motion.

Now suppose (x_o,t_o) is any point in E^{n+1}. Then the surface

$$r(x,t) = {}_E\int_{(x,t)}^{(x_o,t_o)} U(\tau,x,\dot{x}).d\tau = h > 0$$

will be called the <u>elementary sphere of radius h</u> centred at (x_o,t_o). It is clear that $r(x,t)$ satisfies the conditions of Chapter IV, $r_t=-H(t)$, $r_x=q(t)$.

Consider the two surfaces of constant Present Value, illustrated in Figure 10, $V(x,t)=\alpha$ and $V(x,t)=\alpha+h$, where α, $h>0$. Let (x_2,t_2) be any point on the surface $V(x,t)=\alpha$, and let (x_1,t_1) be the point of intersection of the extremal passing through (x_2,t_2) with the surface $V(x,t)=\alpha+h$. Since the two surfaces $V(x,t)=\alpha$, and $V(x,t)=\alpha+h$ are parallel,

$${}_E\int_{(x_1,t_1)}^{(x_2,t_2)} U(\tau,x,\dot{x}).d\tau = h$$

The elementary sphere $r(x,t)$ of radius h centred at (x_2,t_2) thus meets the surface $V(x,t)=\alpha+h$ at (x_1,t_1).

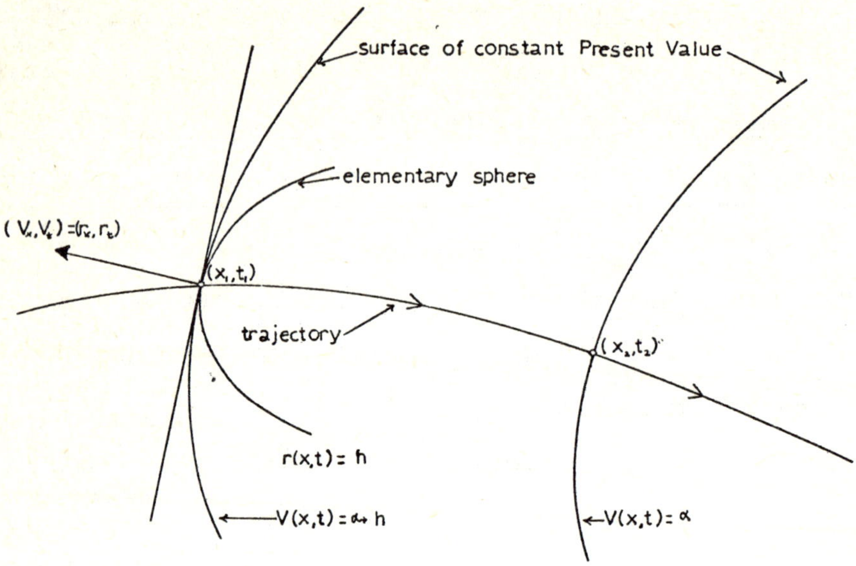

Figure 10. Huygens Principle

Since

$$r_t(x_1, t_1) = -H(t_1, x_1, q_1) = V_t(x_1, t_1)$$

$$r_x(x_1, t_1) = q_1(t_1) = V_x(x_1, t_1)$$

the gradients of the elementary sphere $r(x,t)$ and the surface $V(x,t) = \alpha + h$ coincide at (x_1, t_1). The elementary sphere is thus tangent to the surface $V(x,t) = \alpha + h$ at (x_1, t_1). If, furthermore the objective function $U(t,x,\dot{x})$ is concave in $\dot{x}$, for a maximum problem, or convex in $\dot{x}$ for a minimum problem, then the elementary sphere lies entirely on one side of the surface $V(x,t) = \alpha + h$, as shown in Figure 10, [Rund (1966, 67)]. Since this construction can be carried out for any point centred on the initial surface $V(x,t) = \alpha$ it is clear that the surface $V(x,t) = \alpha + h$ is the envelope of the elementary spheres of radius h centred on the surface $V(x,t) = \alpha$. This is exactly Huygens Principle of propagation of wave-fronts (1678, 68). The surfaces of constant Present Value are thus propagated in the same way as the surfaces of constant phase or the wave-fronts in the theory of light; just as the Present Value surfaces correspond to the wave-fronts, so the trajectories of the economic system correspond to the rays of light.

35. Huygens Principle leads to a more general interpretation of the duality between allocation and valuation. For the duality between the quantities $x_i(t)$ and the imputed values $q_i(t)$, embodied in the dual equations of motion, becomes, through the Hamilton-Jacobi equation, a duality between the rays or the trajectories of the system, and the wave-fronts or surfaces of constant Present Value, the gradients of which are the imputed values.

36. Huygens Principle is equally applicable in analytical mechanics; the surfaces of constant Action of a mechanical system are propagated in exactly the same way as the surfaces of constant Present Value of an economic system, the surfaces of constant Action marking off equal distances, measured in terms of the Lagrangean, $L(t,x,\dot{x})$, along the trajectories of motion. There is thus a striking parallel between the duality of motion in optics, analytical mechanics and economic theory. Perhaps this duality theory can be made to reveal further types of economic motion which have hitherto not been suspected.

It is well known for example, that in optics, the theory of light developed in terms of the rays and wave-fronts, known as geometrical optics, is only able to explain a limited range of phenomena. The phenomena of interference and diffraction for example, cannot be explained within the framework of geometrical optics. For to explain interference and diffraction light must be recognised as involving a wave motion, of which the wave-fronts are the surfaces of constant phase. It is in this way that a transition is made between geometrical optics and wave optics.

Similarly in analytical mechanics, the theory of motion developed in terms of the trajectories and the surfaces of constant Action, $S(x,t) = \alpha$, which correspond to the rays and the wave-fronts in the theory of light, is able to explain only a limited range of phenomena. The phenomena appearing on the atomic and subatomic scale (length less than 10^{-8}cm) such as the diffraction of electrons by crystals, or the quantisation of energy levels, which give rise to the spectra of the elements, for example, cannot be explained within the framework of classical mechanics. To explain these phenomena matter must also be recognised as involving a wave motion, of which the surfaces of constant Action, act as the surfaces of constant phase. It is in this way, as Schrödinger (1926, 69) first showed, that a transition is made between the classical analytical mechanics and quantum mechanics.

The transition from geometrical optics to wave optics and from classical mechanics to quantum mechanics was based on firm empirical evidence: geometrical optics was extended to the more general theory of wave optics only after the phenomena of interference and diffraction of light had been observed, and after the earlier theory had been shown to be incapable of explaining these phenomena. Similarly, classical mechanics was extended to quantum mechanics only after a wealth of empirical evidence on the atomic level had demonstrated the inadequacy of the earlier physical theory. It was empirical evidence that led to the transition in both cases.

It should however be noted that in the evolution of quantum mechanics, empirical evidence did not precede the theory at every stage. For it will be recalled that DE BROGLIE (1925, 70) suggested the idea of associating with a particle of matter of energy E, and momentum p, a wave of frequency $\nu = \frac{E}{h}$, and wave-length $\lambda = \frac{h}{p}$, where h is Planck's constant, <u>before</u> any direct physical evidence had revealed the existence of such a wave. He was thus led to predict the possibility of interference and diffraction of electrons, before these phenomena had ever been observed. As is well known DE BROGLIE'S daring hypothesis was subsequently very beautifully confirmed. Empirical evidence need not therefore always precede the development of the theory.

37. The close parallel between geometrical optics, classical analytical mechanics, and the present economic theory of motion, suggests that the economic theory developed in terms of the trajectories and surfaces of constant Present Value, $V(x,t) = \alpha$, may be extended to a <u>wave theory of motion</u>. The extension is suggested by the remarkable similarity between the innate structure of the theory in each case, for phenomena of this kind do not yet appear to have been observed in economics.

Suppose therefore that as in the transition from classical mechanics to quantum mechanics, the surfaces of constant Present Value, act as the surfaces of constant phase of a wave motion. If in particular the surfaces $\phi(x,t) = \alpha$, denote the surfaces of constant phase of a harmonic wave motion, then the displacement characterising the wave may be written as

$$\Psi(x,t) = k \cos 2\pi\phi(x,t) = \Re(ke^{2\pi i\phi(x,t)}) \qquad (1)$$

where k denotes the amplitude, or maximum displacement, and where $\Re(e^{i\phi})$ denotes the real part of the complex number $e^{i\phi}$. (In the future we will omit $\Re$ in (1) using the convention that Ψ is always the real part of this complex function.) Since $V(x,t)$ has the dimension of the objective function times time, namely the dimension of utility, we need to introduce a <u>constant β</u>, of the same dimension, in such a way that

$$\phi(x,t) = \frac{1}{\beta} . V(x,t) \tag{2}$$

<u>The displacement of the wave associated with the economic system</u> is then

$$\Psi(x,t) = k . e^{\frac{2\pi i}{\beta} V(x,t)}$$

Over a small space-time region the phase function $\phi(x,t)$ can be approximated by the linear terms of its TAYLOR series expansion: without loss of generality this point may be taken as the origin $(0,0)$. Thus

$$\phi(x,t) = \phi(0,0) + x\phi_x(0,0) + t\phi_t(0,0)$$

the wave function in (1) may then be written as

$$\Psi(x,t) = k . e^{2\pi i [x\phi_x + t\phi_t + \phi_o]}$$

Now a plane wave with propagation vector k, and frequency ν has the equation,

$$k . e^{2\pi i [k . x - \nu t + a]} \tag{3}$$

where a is constant. Thus

$$\phi_x = k \ , \ \phi_t = -\nu \ .$$

This together with (2) implies

$$-\nu = \phi_t = \frac{V_t}{\beta} = -\frac{H}{\beta} \qquad (4)$$

$$k = \phi_x = \frac{V_x}{\beta} = \frac{q}{\beta} \qquad (5)$$

These are the <u>fundamental relations</u> for the transition to the wave theory. For (4) implies that we should associate with the economic system a frequency $\nu = \frac{H}{\beta}$, while (5) implies that the economic system is assigned a <u>wave vector</u> $k = \frac{q}{\beta}$.

If n denotes the unit normal to the plane wave (3), and w denotes its velocity of propagation, then

$$k = \frac{\nu}{w} . n$$

thus

$$\|k\| = \frac{\nu}{w} . \|n\| = \frac{\nu}{w}$$

or

$$w = \frac{\nu}{\|k\|}$$

where

$$\|k\| = \sqrt{\sum_{i=1}^{n} k_i^2} \quad .$$

The <u>velocity of propagation</u> of the wave associated with the economic system at any point is thus

$$w = \frac{H}{\|q\|}$$

Since the velocity of propagation w , of a wave is the product of its wave-length λ , and its frequency ν , this implies that the <u>wave-length</u> of the economic system is

$$\lambda = \frac{w}{\nu} = \frac{\beta}{\|q\|}$$

Thus from the relations (4) and (5) we have obtained a complete characterisation of the wave associated with the economic system.

38. Since the wave function for the economic system is

$$\Psi(x,t) = k.e^{\frac{2\pi i}{\beta}V(x,t)}$$

this implies

$$\Psi_t = \Psi.\frac{2\pi i}{\beta}.V_t$$

$$\Psi_x = \Psi.\frac{2\pi i}{\beta}.V_x$$

thus

$$V_t.\Psi = \frac{\beta}{2\pi i}.\Psi_t$$

$$V_x.\Psi = \frac{\beta}{2\pi i}.\Psi_x \qquad (6)$$

In (6) V_t and V_x are to be considered as differential operators acting on the wave function Ψ. Thus with V_t we associate the operator $\frac{\beta}{2\pi i}\frac{\partial}{\partial t}$, while with V_x we associate the operator $\frac{\beta}{2\pi i}\nabla$, where ∇ is the gradient operator $\nabla = (\frac{\partial}{\partial x_1},\ldots,\frac{\partial}{\partial x_n})$. The operator equation

$$\left[V_t + H(t,x,V_x)\right]\Psi(x,t) = 0$$

thus becomes

$$\left[\frac{\beta}{2\pi i}\frac{\partial}{\partial t} + H(t,x,\frac{\beta}{2\pi i}\nabla)\right]\Psi(x,t) = 0$$

which is the fundamental wave equation for the economic system. Just as in the earlier theory the dynamical properties of the system were summarised in its Present Value Function $V(x,t)$, so in the wave theory the dynamical properties of the system are summarised in its wave function $\Psi(x,t)$.

It may perhaps be instructive to consider a purely formal example of this transition to the wave theory. Thus suppose the economic system in which the discrete quantum of utility β appears, has an objective function whose form is similar to that in the earlier examples, except that it is independent of time,

$$U(k,\dot{k}) = \sqrt{g(k) - \dot{k}} \quad , \quad \dot{k} \leq g(k)$$

then its dual objective function is

$$H(k,q) = \frac{1}{4q} + g(k).q$$

Since the total imputed utility is constant during the motion, $H = \alpha$, a solution may be obtained in the form

$$\Psi(k,t) = \psi(k)e^{\frac{2\pi i}{\beta}.\alpha t}$$

where $\psi(k)$ satisfies the time-independent wave equation

$$H\psi = \alpha\psi$$

namely

$$\psi''(k) + \frac{1}{g(k)}\left(g'(k) - \frac{\alpha}{\beta}2\pi i\right)\psi'(k) - \left(\frac{\pi}{\beta}\right)^2 \psi(k) = 0$$

39. Since it seems reasonable to interpret the wave function of an economic system in a manner similar to that in quantum mechanics and wave optics, we are led to interpret the square of the wave function $\Psi(x,t)$ as the probability that the economic system has the co-ordinate value x at time t . This implies that after

normalisation

$$\int_{\mathbb{R}^n} |\Psi(x,t)|^2 . dx = \int_{\mathbb{R}^n} P(x,t) . dx = 1 \qquad (7)$$

where $P(x,t) = |\Psi(x,t)|^2$ is the probability density function characterising the motion in the co-ordinate space. Note that (7) implies that the wave function $\Psi(x,t)$ must be a square-integrable function defined on $\mathbb{R}^n$.

The probability density function for the imputed value space is derived from the FOURIER transform $\Phi(q,t)$ of the wave function $\Psi(x,t)$:

$$\Phi(q,t) = \beta^{-\frac{n}{2}} \int_{\mathbb{R}^n} \Psi(x,t) . e^{-\frac{2\pi i}{\beta} xq} . dx$$

Since the wave function $\Psi(x,t)$ is square-integrable, there is a one-to-one correspondence between $\Psi(x,t)$ and its conjugate function $\Phi(q,t)$; by the FOURIER-PLANCHEREL theorem

$$\Psi(x,t) = \beta^{-\frac{n}{2}} \int_{\mathbb{R}^n} \Phi(q,t) . e^{\frac{2\pi i}{\beta} xq} . dq \quad .$$

Note that PARSEVAL'S equality implies

$$\int_{\mathbb{R}^n} |\Phi(q,t)|^2 . dq = \int_{\mathbb{R}^n} |\Psi(x,t)|^2 . dx = 1$$

$Q(q,t) = |\Phi(q,t)|^2$ is thus a well-defined probability density function on the imputed value space. $Q(q,t)$ denotes the probability that the imputed value is q at time t .

Since $\Phi(q,t)$ and $\Psi(x,t)$ are FOURIER transforms of one another, the form of the function $\Phi(q,t)$ is always related to the form of the function $\Psi(x,t)$. Thus if $\sigma(x_i,t)$ and $\sigma(q_i,t)$

denote the standard deviations of the co-ordinate x_i and its imputed value q_i at time t, then it may be shown that

$$\sigma(x_i,t).\sigma(q_i,t) \geq \frac{\beta}{4\pi} , \quad i = 1,\ldots,n \tag{8}$$

[MESSIAH (1958, 71)]. If the standard deviation is used as a measure of uncertainty, then (8) implies that at any instant of time t, the uncertainty associated with the co-ordinate x_i is inversely related to the uncertainty associated with its imputed value q_i. There is thus always a minimum uncertainty or a maximum certainty that can be attained in predicting the (x_i,q_i) values at any instant, and β determines this maximum certainty. Provided that the present extension of the wave theory to economic theory is valid, this emergence of uncertainty, an exact statement of which is contained in the two probability density functions Ψ and Φ, in a problem that is otherwise apparently deterministic, will appear as one of the fundamental characteristics of the transition to the wave theory of motion.

40. The basic Principle of Motion for the earlier theory may be obtained directly from the wave theory by considering a condition that must be satisfied when the wave-length of the economic system becomes very small. Thus at time zero, every possible path $x(t)$ including the one actually taken by the system, generates a certain Present Value $P(x)$, in such a way that $\frac{P(x)}{\beta}$ denotes the phase of the wave associated with the economic system. Consider an arbitrary path $x(t)$; a neighboring path $\bar{x}(t)$ gives rise to a Present Value $P(\bar{x})$: if the difference $|P(x) - P(\bar{x})|$ between these Present Values is positive, then the phase difference

$$\frac{|P(x) - P(\bar{x})|}{\beta} \tag{9}$$

of the associated waves will be considerable since β is very small. Now it is well-known from the interference theory of waves, that when the phases of two waves differ considerably, the waves interfere in such a way that the displacement of the resulting wave is very small. Only in the region where the phase difference is very small, will a significant displacement arise: in view of the definition of the density function $\Psi(x,t)$ this is therefore the only region in which the actual path of the system is likely to lie. But this implies that (9) must be very small even as $\beta \to 0$. The trajectory of the system thus lies in the region where the variation of the Present Value, for neighboring paths, vanishes to terms of the order β, which is exactly the trajectory implied by the Principle of Stationary Present Value.

The earlier theory is the small wave-length limit of the wave theory: as the wave-length of the economic system becomes very small the economic system need no longer be characterised by its wave function $\Psi(x,t)$, but may be characterised by its phase function $V(x,t)$, for the wave equation reduces to the HAMILTON-JACOBI equation.

* * * * *

BIBLIOGRAPHY

1. Galileo, G., Discorsi e Dimostrazioni Mathematiche Intorno a Due Nuoue Scienze, Leyden, 1638. Also translated as "Dialogues Concerning Two New Sciences" by H. Crew and A. de Salvio, Dover Publications, Inc., New York.

2. Newton, I., Philosophiae Naturalis Principia Mathematica, London, 1687. Also translated as "Mathematical Principles of Natural Philosophy" by A. Motte, University of California Press, Los Angeles, 1962.

3. Euler, L., Methodus Inveniendi Lineas Curvas Maximi Minimive Proprietate Gaudentes, Lausanne, 1744. Reprinted in "Leonhardi Euleri Opera Omnia" Vol 24, Ser. 1 Bern, 1952.

4. Lagrange, L., Mécanique Analytique, Paris, 1788. Also third edition, edited by M. J. Bertrand, published by Mallet-Bachelier, Paris, 1853.

5. Maupertuis, M., "Accord de Differentes Loix de la Nature, qui Avoient Jusqu'ici Paru Incompatibles." Mémoires de l'Académie des Sciences de Paris, 1744.

6. Hamilton, W. R., "On a General Method in Dynamics." Philosophical Transactions of the Royal Society 1834. Also in "The Mathematical Papers of Sir William Rowan Hamilton." Vol. II. Dynamics, Cambridge University Press, 1940.

7. Hamilton, W. R., "Second Essay on a General Method in Dynamics." Philosophical Transactions of the Royal Society 1835. Also in "The Mathematical Papers..." see (6) above.

8. Hamilton, W. R., "Theory of Systems of Rays." Transactions of the Royal Irish Academy 1828. Also in "The Mathematical Papers..." Vol. I Geometrical Optics, see (6) above.

9. Jevons, W. S., The Theory of Political Economy, 1871. Reprinted by A. M. Kelley, New York, 1965.

10. Fisher, I., "Mathematical Investigations in the Theory of Value and Prices." Transactions of the Connecticut Academy, 1892. Reprinted by A. M. Kelley, New York, 1965.

11. Edgeworth, F. Y., Mathematical Psychics, C. Kegan Paul and Co., London, 1881.

12. Hotelling, H., "A General Mathematical Theory of Depreciation." Journal of the American Statistical Association, 1925.

13. Amoroso, L., "The Transformation of Value in the Productive Process." Econometrica, 1940.

14. Samuelson, P. A., "The Stability of Equilibrium: Comparative Statics and Dynamics." Econometrica, 1941. Reproduced in J. Stiglitz (ed.), The Collected Scientific Papers of Paul A. Samuelson, Cambridge, Mass, The M.I.T. Press, 1966, Vol. I.

15. Samuelson, P. A., "The Stability of Equilibrium: Linear and Nonlinear Systems." Econometrica, 1942. See also (14).

16. Samuelson, P. A., Foundations of Economic Analysis, Harvard University Press, Cambridge, 1947.

17. Hicks, J. R., Value and Capital, Oxford University Press, 1939.

18. Malthus, T. R., Principles of Political Economy, 1836. Reprinted by A. M. Kelley, New York, 1968.

19. Rae, J., Statement of Some New Principles on the Subject of Political Economy, 1834. Reprinted by A. M. Kelley, New York, 1964.

20. Mill, J. S., Principles of Political Economy, 1848. Reprinted by A. M. Kelley, New York, 1965.

21. Böhm-Bawerk, E. Von, Positive Theorie des Kapitales, 1889. Translated as "Positive Theory of Capital" Libertarian Press, Illinois, 1959.

22. Wicksell, J. G. K., Vorlesungen über Nationalökonomie, 1901. Translated as "Lectures on Political Economy," reprinted by A. M. Kelley, New York, 1967.

23. Ramsey, F. P., "A Mathematical Theory of Saving." Economic Journal, 1928.

24. Samuelson, P. A., and Solow, R. M., "A Complete Capital Model Involving Heterogeneous Capital Goods." The Quarterly Journal of Economics, 1956. See also (14).

25. Samuelson, P. A., "Efficient Paths of Capital Accumulation in Terms of the Calculus of Variations." K. J. Arrow, S. Karlin, and P. Suppes, (eds.), Mathematical Methods in the Social Sciences, Stanford University Press, 1960. See also (14).

26. Uzawa, H., "Optimal Growth in a Two-Sector Model of Capital Accumulation." Review of Economic Studies, 1964.

27. Scrinivasan, T. N., "Optimal Savings in a Two-Sector Model of Growth." Econometrica, 1964.

28. Samuelson, P. A., "A Catenary Turnpike Theorem Involving Consumption and the Golden Rule." American Economic Review, 1965.

29. Cass, D., "Optimum Growth in an Aggregative Model of Capital Accumulation." Review of Economic Studies, 1965.

30. Cass, D., "Optimum Growth in an Aggregative Model of Capital Accumulation: A Turnpike Theorem." Econometrica, 1966.

31. Koopmans, T. C., "On the Concept of Optimal Economic Growth," in "The Econometric Approach to Development Planning," North-Holland Publ. Co. and Rand McNally, 1966.

32. Malinvand, E., "Croissances optimales dans un modèle macroéconomique," in "the Econometric Approach to Development Planning," North-Holland Publ. Co. and Rand McNally, 1966.

33. Shell, K., (ed.), Essays on the Theory of Optimal Economic Growth, The M.I.T. Press, Cambridge, 1967.

34. Todhunter, R., Institute of Actuaries' Textbook, Part I, Layton, London, 1915.

35. Fisher, I., The Nature of Capital and Income, 1906. Reprinted by A. M. Kelley, New York, 1965.

36. Samuelson, P. A., "Some Aspects of the Pure Theory of Capital." Quarterly Journal of Economics, 1937. See also (14).

37. Neumann, J. Von., "Über ein ökonomisches Gleichungssystem und eine Verallgemeinerung des Brouwerschen Fixpunktsätzes," Ergebnisse eines Mathematischen Kolloquiums, No 8, 1935-36, Leipzig and Vienna, Franz Deuticke, 1937. Also translated as "A Model of General Economic Equilibrium," Review of Economic Studies, 1945.

38. Kantorovitch, L. V., Matematicheskie metody organizatii i planirovania proizvodstva (Mathematical Methods in the Organisation and Planning of Production), Izdanie Leningradskogo Gosudarstvennogo Universiteta, Leningrad, 1939.

39. Dantzig, G. B., "Maximisation of a Linear Function of Variables Subject to Linear Inequalities," in "Activity Analysis of Production and Allocation" (ed.) T. C. Koopmans, J. Wiley and Sons, New York, 1951.

40. Koopmans, T. C., "Analysis of Production as an Efficient Combination of Activities," in "Activity Analysis of Production and Allocation" (ed.) T. C. Koopmans, J. Wiley and Sons, New York, 1951.

41. Minkowski, H., "Theorie der Konvexen Körper, insbesondere Begründung ihres Oberflächenbegriffs," Ges. Abh. Bh 2, B. G. Teubner, Leipzig und Berlin, 1911.

42. Arrow, K. J., "An Extension of the Basic Theorems of Classical Welfare Economics," in J. Neyman (ed.) Proceedings of the Second Berkeley Symposium on Mathematical Statistics and Probability, Berkeley, 1950.

43. Debreu, G., "The Coefficient of Resource Utilisation," Econometrica, 1951.

44. Walras, L., Eléments d'Economie Politique, Lausanne, 1874. Translated as "Elements of Pure Economics" R. Irwin, Illinois, 1954.

45. Pareto, V., Cours d'Economie Politique, Lausanne, 1896-97.

46. Malinvand, E., "Capital Accumulation and Efficient Allocation of Resources," Econometrica, 1953.

47. Pontryagin, L. S., (et alia), The Mathematical Theory of Optimal Processes. Interscience Publishers Inc., 1962.

48. Shell, K., "Optimal Programs of Capital Accumulation for an Economy in Which There is Exogenous Technical Change." Printed in (33).

49. Ryder, H. E., "Optimal Accumulation and Trade in an Open Economy of Moderate Size." Printed in (33).

50. Bellman, R., Dynamic Programming, Princeton University Press, Princeton, New Jersey, 1957.

51. Isaacs, R., Differential Games. John Wiley and Sons, New York, 1965.

52. Alembert, J., d', Traité de Dynamique, Paris, 1743. Reprinted by Gauthier-Villars, Paris, 1921.

53. Arrow, K. J., "Le rôle des valeurs boursières pour la réparation la meilleure des risques," Econometrie, Paris, Centre National de la Recherche Scientifique, 1953.

54. Debreu, G., Theory of Value, Cowles Foundation Monograph 17, J. Wiley and Sons, New York, 1959.

55. Houthakker, H., "Revealed Preference and the Utility Function," Economica, 1950.

56. Valentine, F. A., "The Problem of Lagrange with Differential Inequalities as Added Side Conditions," in Contributions to the Calculus of Variations, 1933-37" University of Chicago Press, Chicago, 1937.

57. Euler, L., "Curvarum maximi minimive proprietate gaudentium inventio nova et facilis," Commentarii Academiae

Scientiarum Petropolitanae 1736. Reprinted in "Leonhardi Euleri Opera Omnia," Vol 25, Ser 1, Bern 1952.

58. Legendre, A-M., "Mémoire sur la Manière de Distinguer les Maxima des Minima dans le Calcul des Variations," Mémoires de l'Académie Royale des Sciences, 1786, Paris.

59. Clebsch, A., "Über die Reduction der Zweiten Variation auf ihre Einfachste Form," Journal für die Reine und Angewandte Mathematik, 1858.

60. Carathéodory, C., "Über die Diskontinuierlichen Lösungen in der Variationsrechnung." Dissertation, Göttingen, 1904 Ges. Math. Schriften Bd 1. C. H. Beck'sche Verlag München, 1954.

61. Poincaré, H., "Mémoire sur les Courbes Définies par une Equation Différentielle," Journal de Mathématiques, 1881.

62. Samuelson, P. A., "Market Mechanisms and Maximisation" 1949, see (14).

63. Fisher, I., The Theory of Interest, 1930. Reprinted by A. M. Kelley, New York, 1965.

64. Lagrange, J. L., "Essai d'une Nouvelle Méthode pour Déterminer les Maxima et les Minima des Formules Intégrales Indéfinies," Miscellanea Taurinensia t.II, 1760-1761. Also in "Oeuvres de Lagrange" t.I Gauthier-Villars, Paris, 1867.

65. Jacobi, C. G., Vorlesungen über Dynamik, 1866. Supplementband Ges. Werke, G. Reimer, Berlin 1884.

66. Lerner, A. P., The Economics of Control, Macmillan, New York, 1944.

67. Rund, H., The Hamilton-Jacobi Theory in the Calculus of Variations, Von Nostrand Co., Princeton, New Jersey, 1966.

68. Huygens, C., Traité de la Lumière, 1690. Also translated by S. P. Thompson as "Treatise on Light", Dover Publications, Inc., New York, 1962.

69. Schrödinger, E., "Quantisierung als Eigenwert problem, II" Annalen der Physik 1926. Translated as "Collected Papers on Wave Mechanics, Blackie and Son, London, 1928.

70. Broglie, L. de, "Recherches sur la théorie des quanta." Annales de Physiques, 1925.

71. Messiah, A., Quantum Mechanics, Vol. I., J. Wiley and Sons, Inc., New York, 1958.

* * * * *

Offsetdruck: Julius Beltz, Weinheim/Bergstr.

Lecture Notes in Operations Research and Mathematical Systems

Vol. 1: H. Bühlmann, H. Loeffel, E. Nievergelt, Einführung in die Theorie und Praxis der Entscheidung bei Unsicherheit. 2. Auflage, IV, 125 Seiten 4°. 1969. DM 12,– / US $ 3.30

Vol. 2: U. N. Bhat, A Study of the Queueing Systems M/G/1 and GI/M/1. VIII, 78 pages. 4°. 1968. DM 8,80 / US $ 2.50

Vol. 3: A. Strauss, An Introduction to Optimal Control Theory. VI, 153 pages. 4°. 1968. DM 14,– / US $ 3.90

Vol. 4: Einführung in die Methode Branch and Bound. Herausgegeben von F. Weinberg. VIII, 159 Seiten. 4°. 1968. DM 14,– / US $ 3.90

Vol. 5: L. Hyvärinen, Information Theory for Systems Engineers. VIII, 205 pages. 4°. 1968. DM 15,20 / US $ 4.20

Vol. 6: H. P. Künzi, O. Müller, E. Nievergelt, Einführungskursus in die dynamische Programmierung. IV, 103 Seiten. 4°. 1968. DM 9,– / US $ 2.50

Vol. 7: W. Popp, Einführung in die Theorie der Lagerhaltung. VI, 173 Seiten. 4°. 1968. DM 14,80 / US $ 4.10

Vol. 8: J. Teghem, J. Loris-Teghem, J. P. Lambotte, Modèles d'Attente M/G/1 et GI/M/1 à Arrivées et Services en Groupes. IV, 53 pages. 4°. 1969. DM 6,– / US $ 1.70

Vol. 9: E. Schultze, Einführung in die mathematischen Grundlagen der Informationstheorie. VI, 116 Seiten. 4°. 1969. DM 10,– / US $ 2.80

Vol. 10: D. Hochstädter, Stochastische Lagerhaltungsmodelle. VI, 269 Seiten. 4°. 1969. DM 18,– / US $ 5.00

Vol. 11/12: Mathematical Systems Theory and Economics. Edited by H. W. Kuhn and G. P. Szegö. VIII, IV, 486 pages. 4°. 1969. DM 34,– / US $ 9.40

Vol. 13: Heuristische Planungsmethoden. Herausgegeben von F. Weinberg und C. A. Zehnder. II, 93 Seiten. 4°. 1969. DM 8,– / US $ 2.20

Vol. 14: Computing Methods in Optimization Problems. Edited by A. V. Balakrishnan. V, 191 pages. 4°. 1969. DM 14,– / US $ 3.90

Vol. 15: Economic Models, Estimation and Risk Programming: Essays in Honor of Gerhard Tintner. Edited by K. A. Fox, G. V. L. Narasimham and J. K. Sengupta. VIII, 461 pages. 4°. 1969. DM 24,– / US $ 6.60

Vol. 16: H. P. Künzi und W. Oettli, Nichtlineare Optimierung: Neuere Verfahren, Bibliographie. IV, 180 Seiten. 4°. 1969. DM 12,– / US $ 3.30

Vol. 17: H. Bauer und K. Neumann, Berechnung optimaler Steuerungen, Maximumprinzip und dynamische Optimierung. VIII, 188 Seiten. 4°. 1969. DM 14,– / US $ 3.90

Vol. 18: M. Wolff, Optimale Instandhaltungspolitiken in einfachen Systemen. V, 143 Seiten. 4°. 1970. DM 12,– / US $ 3.30

Vol. 19: L. Hyvärinen, Mathematical Modeling for Industrial Processes. VI, 122 pages. 4°. 1970. DM 10,– / US $ 2.80

Vol. 20: G. Uebe, Optimale Fahrpläne. IX, 161 Seiten. 4°. 1970. DM 12,– / US $ 3.30

Vol. 21: Th. Liebling, Graphentheorie in Planungs- und Tourenproblemen am Beispiel des städtischen Straßendienstes. IX, 118 Seiten. 4°. 1970. DM 12,– / US $ 3.30

Vol. 22: W. Eichhorn, Theorie der homogenen Produktionsfunktion. VIII, 119 Seiten. 4°. 1970. DM 12,– / US $ 3.30

Vol. 23: A. Ghosal, Some Aspects of Queueing and Storage Systems. IV, 93 pages. 4°. 1970. DM 10,– / US $ 2.80